P9-DVP-739

TIME

GLOBAL WARMING

The Causes ▪ The Perils ▪ The Solutions

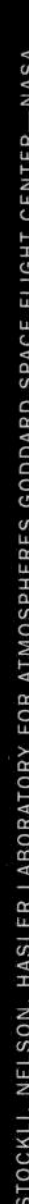

Detail from Oil Fields #19A, Beldrige, Calif., 2003: The oil produced by fields like this one in California helps power the economy, but

it also contributes to climate change and pollution.

TIME

Managing Editor Richard Stengel
Design Director D.W. Pine
Director of Photography Kira Pollack

Global Warming

The Causes ▪ The Perils ▪ The Solutions

Writer and Editor Bryan Walsh
Designer Sharon Okamoto
Photo Editor Dot McMahon
Head Reporter David Bjerklie
Copy Editor Shelley Wolson
Editorial Production David Sloan
Graphics Editor Lon Tweeten

TIME HOME ENTERTAINMENT
Publisher Jim Childs
Vice President, Business Development and Strategy Steven Sandonato
Executive Director, Marketing Services Carol Pittard
Executive Director, Retail and Special Sales Tom Mifsud
Executive Publishing Director Joy Butts
Director, Bookazine Development and Marketing Laura Adam
Finance Director Glenn Buonocore
Assistant General Counsel Helen Wan
Assistant Director, Special Sales Ilene Schreider
Book Production Manager Suzanne Janso
Design and Prepress Manager Anne-Michelle Gallero
Brand Manager Michela Wilde
Associate Brand Manager Isata Yansaneh
Associate Prepress Manager Alex Voznesenskiy
Associate Production Manager Kimberly Marshall

Editorial Director Stephen Koepp
Editorial Operations Director Michael Q. Bullerdick

Special Thanks to:
Christine Austin, Jeremy Biloon, Susan Chodakiewicz, Rose Cirrincione, Jacqueline Fitzgerald, Hillary Hirsch, Christine Font, Jenna Goldberg, Lauren Hall Clark, Amy Mangus, Robert Marasco, Amy Migliaccio, Nina Mistry, Tara Rice, Dave Rozzelle, Adriana Tierno, Vanessa Wu, TIME Imaging

Published by TIME Books, an imprint of Time Home Entertainment Inc.
135 West 50th Street • New York, N.Y. 10020

ISBN 10: 1-60320-248-X, ISBN 13: 978-1-60320-248-0 LOC: 2012938722

We welcome your comments and suggestions about TIME Books. Please write to us at: TIME Books, Attention: Book Editors, P.O. Box 11016, Des Moines, IA 50336-1016

If you would like to order any of our hardcover Collector's Edition books, please call us at 1-800-327-6388, Monday through Friday, 7 a.m. to 8 p.m., or Saturday, 7 a.m. to 6 p.m., Central Time.

Cover: Sue Flood—The Image Bank/Getty Images
Back Cover: Jon Hicks—Corbis

Contents

Wind turbines coat the land around the California desert town of Palm Spring.

JON HICKS—CORBIS

How Hot Does It Have To Get?

By Bryan Walsh

At the first Earth Summit, held in 1992, leaders from around the world—including U.S. President George H. W. Bush—converged in Rio de Janeiro to show concern about the imperiled global environment. But mostly they paid lip service. For all the international power on display, it was a speech by a 12-year-old girl named Severn Suzuki that truly captured the moment. "I am only a child," she told the assembled audience. "Yet I know that if all the money spent on war was spent on ending poverty and finding environmental answers, what a wonderful place this world would be ... You grownups say you love us, but I challenge you, please, to make your actions reflect your words."

Twenty years later, Suzuki was back in Rio, this time for the second Earth Summit in June 2012. She was older, wiser—and sadder. In interviews, Suzuki noted that for all the talk in Rio—and in all the climate and environmental summits that have followed it—we're losing ground in the war to save the environment. A report released before the second Earth Summit by the U.N. Environment Programme noted that of the 90 most important international commitments to sustainability, the world has made significant progress on only four. An assessment published the same month in the journal *Nature* warned that because of human activity, the planet was reaching a potentially catastrophic tipping point. Twenty percent of vertebrate species are at risk of extinction, coral reefs have declined by 38% since 1980, and greenhouse-gas emissions could increase 50% by 2050. As Suzuki told a reporter in Rio in 2012, it "is clear that we have not achieved the sustainable world we knew we needed 20 years ago."

While I haven't been following environmental issues for quite as long as Suzuki has, I've seen that same cycle of hope and disappointment that seems to drive our efforts to save the planet. When I began my work as Time's environment writer in 2007, interest in global warming and the environment was at a high. Al Gore had just won an Oscar for the documentary *An Inconvenient Truth* and shared a Nobel

---

Bryan Walsh *is a senior editor for TIME International and the environment writer for TIME magazine. He created and writes the Ecocentric blog on TIME.com.*

DOCE—REUTERS

At the Rio 20 U.N. Sustainable Development Summit in June 2012, a banner echoed the theme of two decades earlier.

Peace Prize for his work on climate change. After years of paralysis under President George W. Bush, the U.S. seemed ready to join the rest of the world in acting on global warming. In the 2008 Presidential election, both the Democratic and Republican candidates embraced proposed legislation to cut carbon emissions. Global corporations built advertising campaigns around their environmental efforts. It was a good time to be green—or report on it.

Five years later, we seem to have gone backwards. Global warming has become a politically divisive issue in the U.S., where Republicans have turned wholesale against any climate action, as well as internationally. The global energy picture is changing, with new supplies of oil and natural gas—including in the U.S.—complicating the promised clean-tech revolution. Yet even as we dither, the planet keeps on getting hotter, with natural disasters piling up and 2012 on track to be the warmest year on record. The worse things get, the less we seem to be able to do about it, as even the most pressing environmental problems are swallowed by the constant crises in the global economy or security. We can't seem to be bothered to save ourselves.

Yet the reality is more complex than that. If the case for climate action has been dented and bruised over the past few years, it reflects a growing realization of just how difficult it will be to halt the warming of the planet without compromising the economic growth on which we all depend. We may miss the certainty we felt five years ago about the crusade to save the planet—let alone 20 years ago at the Earth Summit—but we also should have realized that this was never going to be easy.

Still, as the chapters of this book show, if there's much to fear, there's also a basis for hope. Activists in the field and scientists in the lab are pushing for solutions to our most pressing environmental threats. New technologies like advanced solar and fuel cells promise sources of energy beyond fossil fuels. Even everyday actions like biking to work or air-drying your laundry can add up.

As Severn Suzuki knows, it's hard to keep hoping when you've been disappointed so many times in the past. But we have to believe that this time will be different, if only because we have no other choice, and no other planet to call home.

Planet in Peril

The Earth has been through a lot in its 4.5 billion years of existence, but human beings are its biggest challenge yet. An explosively growing human population, already 7 billion plus, is using up natural resources, stripping the land of forests, and polluting the air and water. Most of all we're adding billions of tons of greenhouse gases to the atmosphere, changing the relatively stable climate humans have depended on for tens of thousands of years. That could mean devastating floods, scorching droughts, and mega-storms that could wreck economies and cost lives. It's tempting to dismiss the fears of environmentalists, but it's not just the planet that will be in peril if we fail to deal with climate change. It's all of us—and generations to come.

HAKAN LUDWIGSON

Arctic Melting

If you want to witness climate change in action, just head north. While the world as a whole has warmed about 1°F over the entire 20th century, parts of the Arctic have heated up by 4°F to 5°F just since 1950. In 2011, Arctic sea ice shrank to a summer low 1 million square miles below the 1979–2000 average low (1 million square miles is an area of ice greater than all the U.S. states east of the Mississippi). Some scientists worry that Arctic sea ice may be going into a "death spiral," unstoppable even if we manage to reduce carbon emissions. And that has consequences for the rest of us—what happens in the Arctic doesn't stay in the Arctic.

9529
4
TH COACH
2
GO AHEAD.
PLAN.
AXIS TRIPLE
ADVANTAGE FUND

RANDY OLSON

Population Growth

The 7 billionth person officially came into the world on October 31, 2011. The fact that it was Halloween was coincidental—U.N. demographers estimated the date—but it was appropriate. Population growth has people scared. The sheer number of bodies on the planet is stressing out natural resources and generating more pollution and greenhouse gases. While the good news is that we still have more than enough space and food to support a huge global population, humanity is crowding out other species on the planet, accelerating the rate of extinction. The future will only be more crowded—demographers estimate the world could have 9 billion people by 2050, up 50% in a half-century.

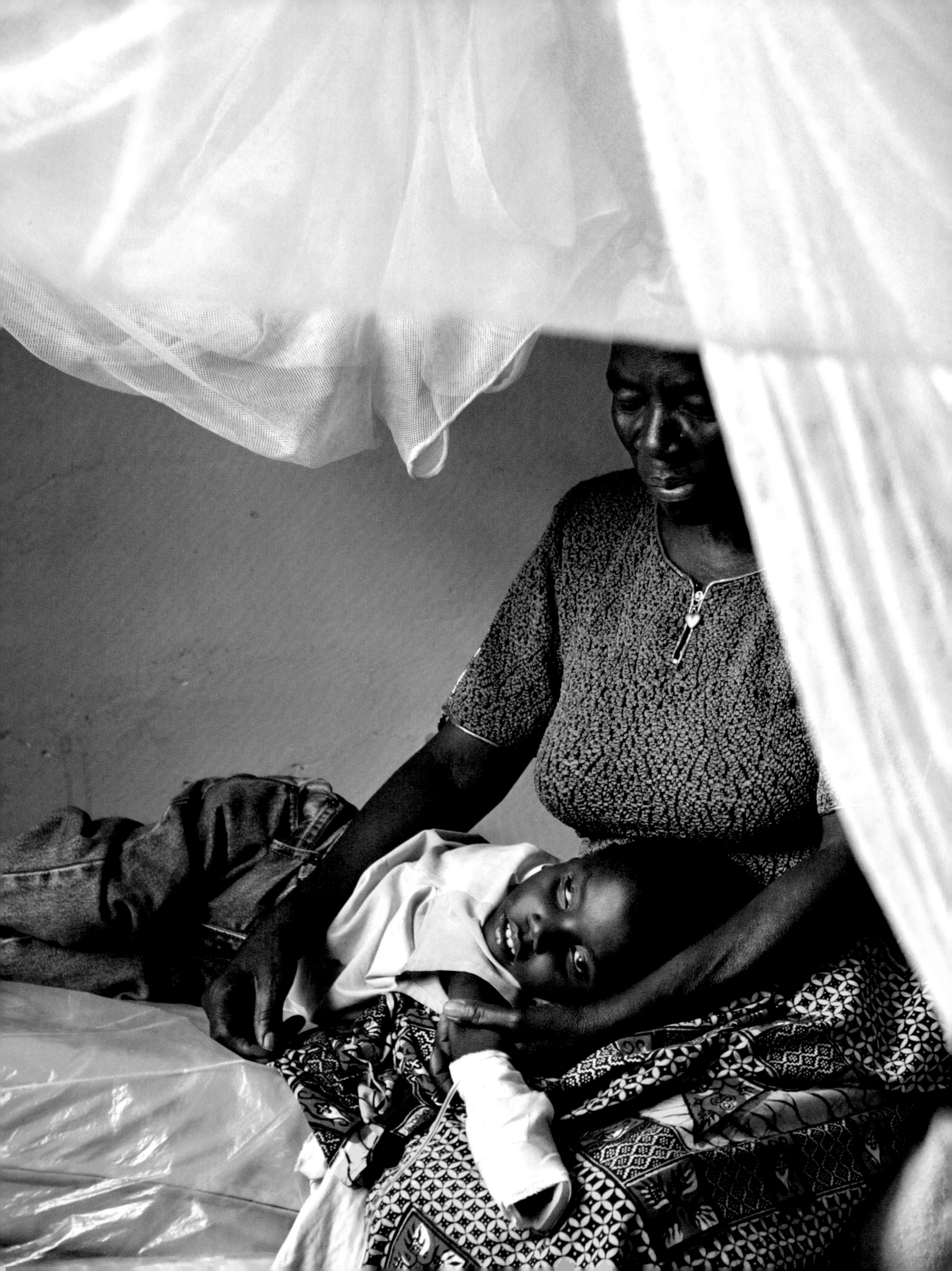

JOHN STANMEYER–VII

Tropical Disease

There's a reason that countries in the tropics tend to suffer disproportionately from deadly infectious disease. The heat and humidity makes it easier for many pathogens to survive and thrive—especially those borne by parasites and insects. That includes malaria, which kills 1 million people a year and afflicts as many as 1 billion people in 109 countries throughout Africa, Asia, and Latin America. As global temperatures rise, the heat will likely enhance the ability of infected mosquitoes to transmit the disease and widen the geographical distribution of the bugs. Climate change won't be the only factor affecting the spread of malaria—far more important than temperature is the ability of governments to control the disease, which is why the rich tropical country of Singapore has all but wiped out malaria. But a warmer world will be one in which malaria and other tropical diseases are an even greater threat to the world's most vulnerable people.

GEORGE STEINMETZ

Drought

Other natural disasters strike suddenly and leave devastation behind. Weather experts describe drought as the "creeping disaster." Though it destroys no homes and yields no direct death toll, drought can cost billions of dollars, lasting for months and even years. A major drought took its toll over much of the Southern and Western U.S. in 2011 and 2012, igniting massive wildfires in Colorado and Montana, ruining crops, and drying up riverbeds and creeks in Texas, such as this one feeding Lake Travis which receded to 26 feet below its normal level in 2011. Years of unusually dry weather have reduced the flow of vital waterways like the Colorado River, even as growing development stresses the remaining water supplies. As the world's climate warms, expect droughts to worsen—most climate models predict that dry areas will get ever drier, which could spell disaster for already arid places like the rapidly growing American Southwest. Even more tragic could be the fate of regions like sub-Saharan Africa, where poverty and drought are a recipe for human catastrophe.

Floods

If drought is the biggest water problem posed by climate change, floods are a close second. A hotter atmosphere can hold more moisture, so as the climate warms, more water will end up circulating above us. Storm systems become supersized, and when the rain comes it falls in torrents. The result can be damaging floods like the one seen here in Bangkok, where weeks of downpours in the fall of 2011 swamped much of Thailand. More than 800 people died in the resulting floods, which cost the country more than $45 billion. Expect those numbers to increase in the future, as populations grow in coastal cities that are prone to floods, putting more and more people and property in harm's way.

NUEL BALIEL—AFP/GETTY IMAGES

SUSAN SCHULMAN—GETTY IMAGES

Deforestation

Environmental news isn't always bad news. Since 2005 the percentage of the world's tropical forests that are under some form of sustainable management has increased 50%, from 69 million hectares to 183 million hectares. Thanks to the work of environmental activists, the pace of deforestation has slowed in much of the world. But that doesn't mean it has stopped. Less than 10% of tropical forests are sustainably managed, and the pressure to clear land for agriculture or settlements in countries like Brazil is only going to grow. That's the bad news for the climate—forests, especially in the tropics, store massive amounts of carbon. When those trees are cut down or burnt, much of the carbon ends up in the atmosphere, which is why deforestation accounts for as much as 15% of global carbon emissions. Add in the fact that tropical forests provide vital habitat for wildlife, and it's clear why we need to save the trees if we want to save the planet.

RIA NOVOSTI/MIKHAIL FOMICHEV—CAMERA PRESS/REDUX

Heat Wave

There's a lot of scientific debate on exactly how climate change will affect the planet. But here's something that we know for sure: It will get hotter. In fact, it already has. Record highs are now set far more frequently than record lows, and 2012 in particular is shaping up to be the warmest year on record. The terrible temperatures take a human toll: A prolonged heat wave in Russia in 2010 killed nearly 11,000 people in Moscow alone, and led to devastating wildfires throughout much of Siberia. Extreme heat also worsens air quality—see the Muscovite here wearing a gas mask during the city's heat wave—and disproportionately kills the very old and the very poor. The good news is that there are ways for cities to adapt to extreme heat: In Chicago, where more than 700 people died during the great heat wave of 1995, city officials have planted trees to moderate temperatures and set up cool zones for elderly residents. Other cities are following suit, but human beings can only stand so much heat.

Climate Change: The Big Questions

Politicians and parts of the public may think climate change is still up for debate, but for nearly every scientist who knows the subject, the case is closed. Greenhouse gases like carbon dioxide are emitted when we burn fossil fuels like coal or oil, and those gases collect in the atmosphere and warm the planet. Deforestation and some natural sources like methane from animals also accelerate climate change as well, but most of the 1°F the world has warmed over the past century is due to man-made causes. In other words, it's our fault—and our responsibility to do something about it.

But just because the basics of climate science have been established doesn't mean we know everything about how fast the planet will warm—and what will happen when it does. Climate modeling is incredibly complex, as scientists use sophisticated computer programs to try to predict how every part of the planet will respond to changing greenhouse gas emissions. That knowledge is vital—better climate science can help identify the local impacts of global warming, and prepare us for exactly what's to come. That's why scientists are working hard on some of the big unanswered questions of climate change:

What Will Climate Change Do to Extreme Weather?

Maybe we should retire the term "global warming," which makes climate change sound like a nice bath. It's true that climate change—caused chiefly by the rapid increase in man-made carbon emissions—will result in warmer temperatures, fewer cold days, and longer and more intense heat waves. But the real damage, both economically and in human lives, is likely to be inflicted by an increase in extreme weather events—floods, storms, droughts.

Arctic sea ice is already vanishing fast—and that's bad news for the climate.

The problem is that actually attributing extreme events to climate change has always been challenging, which makes it more difficult to predict how weather will respond to warming. But scientists are getting better at it, and a report released in 2011 by the U.N.'s Intergovernmental Panel on Climate Change has a clear message: More carbon emissions will mean more dangerous extreme weather events. "We need to be worried," says Maarten van Aalst, the director of the International Red Cross/Red Crescent Climate Centre. "Risk has already increased dramatically."

Exactly how that risk will increase isn't as certain, though. Heat waves will certainly increase, but while scientists believe that tropical storms like hurricanes may get stronger, there's no clear evidence that they will become more frequent. Researchers will continue trying to pin down the relationship between climate change and extreme weather, but adaptation will be as important as prediction.

Will the Melting Arctic Cause Climate Change Feedback?

On the basics, climate science is

CLOCKWISE FROM TOP LEFT: PATRICK PLEUL—EPA/CORBIS; JOHN BAZEMORE—AP; BRIAN J. SKERRY—NATIONAL GEOGRAPHIC SOCIETY/CORBIS; PAUL SOUDERS—CORBIS pretty straightforward. Carbon dioxide released into the air adds to the greenhouse effect, which traps more solar energy in the atmosphere and warms the planet. That's the simple story. But there are wild cards in the climate system, some of which—if they flip the wrong way—could vastly accelerate global warming well beyond anything most climate models predict.

One of those wild cards is the estimated 1,672 billion tons of carbon equivalent trapped in the form of methane in the Arctic permafrost, the soils kept frozen by the far North's extreme temperatures. Methane is a powerful greenhouse gas—it has 20 times the warming effect of carbon dioxide—and the total amount of carbon equivalent in the Arctic permafrost is 250 times greater than annual U.S. greenhouse gas emissions. As the Arctic warms—which it's doing rather rapidly—there's a risk that the permafrost could become less than permanent, releasing some of that trapped methane into the air, which would then accelerate warming, leading to more Arctic melt, more methane emissions...so on and so on. Climate scientists call this a "feedback loop"—and if it happens soon, we could be in deep trouble.

Will the Oceans Become Too Acidic?

Human beings are doing unprecedented things to the Earth, which is somewhat impressive when you realize that the planet has existed for more than 4.5 billion years. But that's what happens when you add tens of billions of tons of carbon dioxide into the atmosphere—and into the oceans as well.

We don't often think about the oceans when we consider climate change. But a quarter of the carbon emitted into the atmosphere is actually absorbed by the oceans. And over time, that carbon is making the oceans more acidic. Over the past century, the ocean pH—which measures the relative acidity of a liquid—has fallen by 0.1 unit to 8.1. That doesn't sound like much, but ocean acidification is happening faster now than it has for at least 300 million years. As the rate of man-made carbon emissions increases, so will the rate of acidification.

Why does that matter?

More acidic waters are bad news for sea creatures that have carbonate shells, which can simply dissolve if pH levels fall too low. That includes coral reefs—the rainforests of the ocean—which provide vital habitats to all sorts of sea creatures. The oceans are changing fast because of climate change—and we have no idea what's coming next.

What's the "Safe" Level for Carbon Dioxide?

Everyone knows that we need to reduce carbon emissions—and fast—to prevent climate change from getting out of hand. But exactly how much do we need to reduce those emissions—and how fast? For a long time scientists believed that the red line for carbon concentration in the atmosphere was 450 parts per million (ppm). To put that in perspective, the carbon concentration in the atmosphere before the Industrial Revolution—when we started burning carbon-heavy fuels like coal and oil—was 280 ppm. The level is currently 396 ppm, and it increases by about 2 ppm to 3 ppm each year. If we can level off carbon emissions in the near future, we should be able to keep atmospheric carbon levels at no more than 450 ppm, which should in turn keep the climate from warming more than 3.6°F. Any hotter than that, and things could start to get seriously dicey.

But at least one scientist, NASA's James Hansen, thinks 450 ppm is far too high. In a 2007 speech at the annual meeting of the American Geophysical Union, Hansen argued that the safe limit for carbon was actually 350 ppm, which means we're already in the danger zone. Hansen noted that even at the current level of carbon, we're already seeing extreme effects, from melting Arctic sea ice to serious heat waves. Returning atmospheric carbon to 350 ppm will be incredibly difficult, but as Hansen himself has written: "The stakes, for all life on the planet, surpass that of any other crisis." In other words, we don't have any choice.

Climate change could lead to serious ocean acidification.

Extreme events like floods could become worse in the future.

Fossil Fuels

Coal, oil, and natural gas power the global economy—and they're contributing to global warming. As we grapple with climate change, we'll need to grapple with our dependence on fossil fuels, even as new sources of oil and natural gas continue to tempt us.

HENRIK WEIS—GETTY IMAGES

The Future Of Oil

Extreme oil–from the deep Atlantic to the Arctic, from fracking in the U.S. to tar sands in Canada–is replacing dwindling supplies and extending the age of petroleum. But it comes at a heavy economic and environmental cost we may not be able to bear.

In the Canadian province of Alberta, energy companies are tapping the enormous reserves of oil sands. But mining the sands can be extremely polluting.

THE WATERS OF THE ATLANTIC Ocean 180 miles east of Rio de Janeiro are a cobalt blue that appears bottomless. But it only seems that way. Some 7,000 feet beneath the choppy surface lies the silent sea floor, and below that is 5,000 feet of salt rock, deposited when the continents of South America and Africa went their separate ways 160 million years ago.

Underneath it all is oil. By one count, the pre-salt reservoirs off the central coast of Brazil hold as much as 100 billion barrels of crude; that's another Kuwait. It's why former Brazilian President Luiz Inácio Lula da Silva called the pre-salt finds a "gift of God," and it's why the massive Cidade de Angra dos Reis floating oil-production facility—operated by Petrobras, Brazil's state-run oil giant—is anchored in the Atlantic, pumping 68,000 barrels of crude a day from one of the deepest wells in the world. The platform deck is so big you could play the Super Bowl on it, if not for the nest of interlocking pipes and valves that circulate oil, methane, and steam throughout the ship. As I tour the deck in an orange safety jumper, a Petrobras engineer named Humberto Americano Romanus urges me to put a hand to one of the oil pipes. I can feel it pulse like an artery, the oil still warm from the deep heat of the earth. "It's 50 barrels a minute passing through here," he says over the din of the platform. "That's a lot of oil."

But not enough. Demand for oil is still rising—set to grow 800,000 barrels a day in 2012 despite a still sluggish global economy. Meanwhile, production from places like Russia, Iran, and Kuwait seems to be plateauing. The rigs that have gathered along the coast of Brazil are drilling deeper than ever before, through layers of salt rock, in some of the most complex and risky operations the industry has ever seen. "This reservoir is very heterogeneous, very challenging," says Jose Roberto Fagundes Netto, general manager of research and development for Petrobras. "But we know an accident is unacceptable." A well blowout like the one that caused the BP oil spill in 2010 would be even harder to contain in the deeper pre-salt waters.

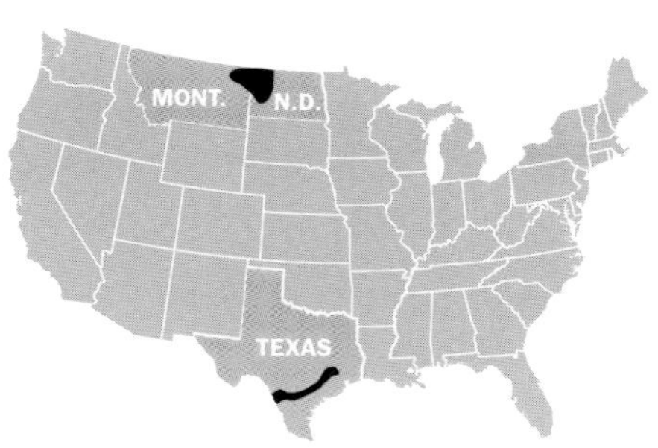

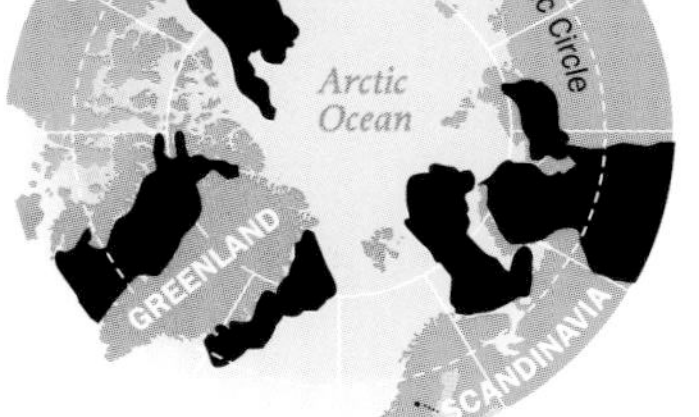

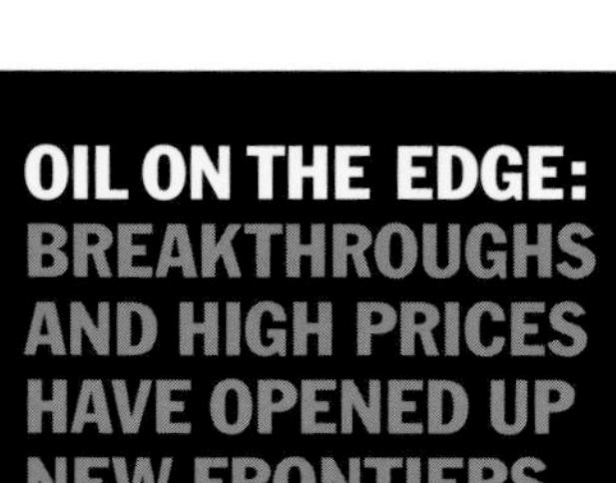

OIL ON THE EDGE: BREAKTHROUGHS AND HIGH PRICES HAVE OPENED UP NEW FRONTIERS FOR PETROLEUM

TIGHT OIL

BACKGROUND
Light crude oil that is bound tightly in **formations of relatively permeable shale.** Wells are drilled vertically and then horizontally into the shale layer. Hydraulic fracturing is used to break the rock underground, and oil flows up the well.

ENVIRONMENTAL IMPACT
Tight oil requires fracking, which involves **injecting millions of gallons of water mixed with chemicals deep into the ground.** There can be a risk of contamination to groundwater, though there have been no proven cases yet. Burning of excess methane from tight-oil wells can cause air pollution.

RESERVES
Up to 300 billion barrels globally.

COST OF PRODUCTION **$50** per barrel.

ARCTIC OFFSHORE

BACKGROUND
As climate change melts Arctic sea ice, **vast areas of water that were once blocked are now opening for offshore drilling and oil shipping.** Call it the unexpected climate-change dividend.

ENVIRONMENTAL IMPACT
Even though the sea ice is melting, Arctic waters remain incredibly treacherous, with icebergs and storms threatening drilling ships. **Any oil spill would be much harder to clean up in the freezing water than in the warmth of the Gulf,** and the remoteness of the Arctic means that it would be difficult to stage a massive spill response.

RESERVES
Estimated **90 billion** barrels.

COST OF PRODUCTION
Unclear but likely **above $100** a barrel.

GRAPHICS BY LON TWEETEN

This is the new world of extreme oil. Petrobras can afford to push the frontiers of offshore drilling because the price of Brent crude, a benchmark used by oil markets, in 2011 averaged $111, the highest average cost since the Drake well in Titusville, Pa., began spewing wealth in 1859, launching the petroleum era. From that time on, even despite J.D. Rockefeller's attempt to monopolize it, oil has experienced a 150-year price slide, interrupted by periodic spikes. The prices of all commodities fluctuate, but oil's irreplaceability—it's the fuel that makes us go—ensures that those spikes hurt. In 2011 oil soared in part because of geopolitics, especially the threat that Iran would block the Gulf of Hormuz and cut supplies. That uncertainty contributed to a risk premium of perhaps $20 or more a barrel. A promise by Saudi Arabia in late March to bring spare oil production onto the market has done little to calm prices.

In the U.S., consumers face an extreme-oil paradox. We need more oil to achieve energy independence—and we're producing it in places like the Bakken shale formation in North Dakota—even as we are using less of it. A combination of recession, conservation, and improved auto efficiency has helped the U.S. shed demand impressively. But demand in China, India, and other developing nations has replaced it. Result: plentiful but expensive oil that translates into painfully high gas prices. In 2011 the average cost for a gallon of unleaded was $3.51, the highest on record, up from $2.90 a year before. That takes a chunk out of household budgets and threatens an already underwhelming economic recovery.

Not that long ago, the big worry about fossil fuels was how rapidly supplies were waning. Now new and unconventional sources of oil are filling the gaps. Ultra-deepwater reserves like those found off Brazil offer the promise of billions of barrels. Technological breakthroughs have unlocked what's known as tight oil in the shale rock of North Dakota and Texas, reversing what seemed like a terminal decline in U.S. oil production. Alberta's vast oil sands have given Canada the world's second largest crude reserves, after Saudi Arabia's, and offer the U.S. a friendlier crude dealer. As global warming melts the Arctic sea ice, an unexpected dividend is access to tens of billions of barrels of oil in the waters of the far north. "We've seen a paradigm shift over the past

PRE-SALT DEEPWATER

BACKGROUND
Reservoirs of oil found **below thick layers of salt beneath the ocean floor that were deposited more than 150 million years ago.** Requires offshore drilling through as much as 9,000 ft. of water, additional rock and more than 5,000 ft. of salt.

ENVIRONMENTAL IMPACT
The pre-salt reservoirs represent some of the most technologically challenging offshore drilling. The wells are deeper than the Gulf of Mexico well that led to the BP oil spill in 2010. **A blowout would be incredibly difficult to control.**

RESERVES
50 billion to 100 billion barrels.

COST OF PRODUCTION
$45 to $65 a barrel.

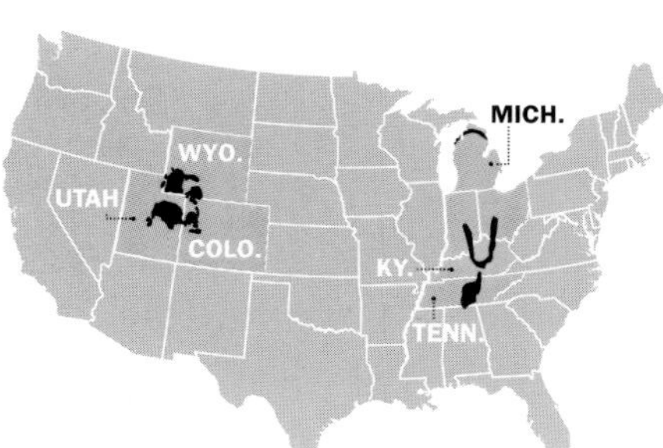

OIL SHALE

BACKGROUND
Shale that contains **a solid bituminous material called kerogen.** The rock has to be mined and then heated to a high temperature to separate the oil from the shale.

ENVIRONMENTAL IMPACT
The cost of mining and processing oil shale is still too high to make the process worthwhile. Oil shale requires significant amounts of land and water and produces toxic tailings. Oil-shale crude also has a larger greenhouse-gas footprint than conventional oil.

RESERVES
800 billion barrels, though estimates remain uncertain.

COST OF PRODUCTION
Over $100 a barrel.

OIL SANDS

BACKGROUND
Loose sand or sandstone that's saturated with **a dense and viscous form of petroleum called bitumen.** The oil sands are exploited either through vast open-pit mines or through in situ wells that process the bitumen underground.

ENVIRONMENTAL IMPACT
Open-pit oil-sands mines **leave large piles of toxic tailings that can pollute nearby water sources.** Gasoline from oil sands results in **10% to 15% more greenhouse-gas emissions per barrel than conventional oil** because of the additional energy needed to refine it.

RESERVES
169 billion recoverable barrels.

COST OF PRODUCTION
$50 to $75 per barrel.

decade," says Daniel Yergin, chairman of the research group IHS CERA. "You look at tight oil and oil sands and deepwater, and you see the results."

Those results could be the problem. While unconventional sources promise to keep the supply of oil flowing, it won't flow as easily as it did for most of the 20th century. The new supplies are for the most part more expensive than traditional oil from places like the Middle East, sometimes significantly so. They are often dirtier, with higher risks of accidents. The decline of major conventional oil fields and the rise in demand mean the spare production capacity that once cushioned prices could be gone, ushering in an era of volatile market swings. And burning all this leftover oil could lock the world into dangerous climate change.

Just how dangerous? Oil is responsible for about 40% of the carbon emissions from fossil fuels worldwide, so continued dependence on crude will make it that much more difficult to deal with climate change. And some of those new sources of crude are particularly dirty, including the much-hyped Canadian oil sands, which can produce as much as 15% to 20% more greenhouse gases than conventional oil. "I'm less concerned about the environmental consequences of pursuing what's left," says Michael Klare, an energy expert and the author of *The Race for What's Left.* There will be oil, but it will be expensive, dirty, and dangerous.

The Bakken Boom

If you want to find oil in the U.S., or a job, for that matter, head to North Dakota. The Peace Garden State is experiencing a remarkable oil boom in the midst of high gas prices, with production rising from 98,000 barrels a day in 2005 to more than 510,000 barrels by the end of 2011. Thanks to shale oil in the Bakken formation, the petroleum workforce has risen from 5,000 in 2005

CLOCKWISE FROM LEFT: DANIEL BELTRÁ FOR GREENPEACE; STEVE MORGAN—GREENPEACE; NICOLE BENGIVENO—THE NEW YORK TIMES/REDUX

The consequences of our addiction to crude can be seen in the Gulf oil spill, left. A similar disaster in the Arctic, top right, would be worse, while North Dakota is growing as a new supplier, bottom right.

to more than 30,000 people. North Dakota's unemployment rate is the nation's lowest, 3.2%, and so many would-be roughnecks have flooded the state that workers are housed in temporary "man camps," like Wild West mining settlements. "You can go straight to those fields and get a good-paying job," says Scott Tinker, the state geologist of Texas. "The demand is there."

So is the supply, thanks to innovations in hydraulic fracturing and horizontal drilling that have opened up reserves of oil previously considered unobtainable. Using a process similar to one employed in shale-gas exploration, which has flooded the U.S. with cheap natural gas, rigs drill down first and then horizontally into shale layers before fracturing the rock to release the tightly bound oil. "The same massive investment we saw with shale gas is now happening with tight oil," says Seth Kleinman, an analyst with Citigroup who recently wrote a research note on the potential of tight oil. "And it's going to play out in the same massive way."

Tight oil has helped revitalize the American drilling industry and it could contribute significantly to global supplies, with the International Energy Agency (IEA) projecting that U.S. tight-oil production could reach 2.4 million barrels a day by 2020.

Thanks as well to greater efficiency. In 2011 the U.S. imported just 45% of the liquid fuels it used, down from a peak of 60% in 2005, and just 1.8 million barrels a day came from the Persian Gulf. If domestic oil production continues to rise, the U.S. could actually approach a goal that has long seemed a political fantasy: energy independence.

But just how much more the U.S. will be able to produce is up for debate. While tight-oil reserves are plentiful, wells tend to dry up quickly, which means a lot of drilling is needed to keep the oil flowing. Even if the U.S. can't achieve energy independence, the oil-sand

resources of Canada, already America's biggest oil supplier, could further reduce imports from the Middle East. High oil prices have boosted investment in the oil sands, and the Energy Information Administration (EIA), the analytical arm of the U.S. Energy Department, projects that oil-sand production will rise from 1.7 million barrels a day in 2009 to 4.8 million barrels in 2035—more than Iran's current output.

So does that mean the return of $2-a-gallon gasoline? Nope. It's true that reducing oil imports is good for the U.S. economy. Americans spent $331.6 billion—the size of the entire agriculture industry—on oil imports last year, up 32% from 2010. Cutting imports keeps that money in the U.S., reducing a trade deficit that hit $560 billion last year. It's also, of course, good for international oil companies like Shell and Chevron, which are increasingly being squeezed out by massive state-owned companies. You may not like Exxon because of the pump price or its oversize profits, but how much love do you have for autocratic petrostates like Iran or Russia? Exxon's growth trickles down; the oil-and-gas industry created 9% of all new jobs last year, according to a report by the World Economic Forum, even as oil companies booked multibillion-dollar profits.

But contrary to what the drill-here, drill-now crowd says, oil companies could punch holes in every state and barely make a dent in gasoline prices. Even a more energy independent U.S. can't control prices, not with a thirsty China competing on the globalized oil market. "Energy security is fine, but it doesn't have that much meaning in a globalized economy," says Guy Caruso, a former head of the EIA. "More production adds fungibility to the world market, but we're still vulnerable to shocks in other countries." The oil the U.S. uses may be American, but that doesn't mean it will be cheap.

The True Price of Oil

Then there's the environmental cost. Oil has never exactly been clean, but the new sources coming online tend to be more polluting and more dangerous than conventional crude. Producing oil from the sands in northern Alberta can be destructive to the local environment, requiring massive open-pit mines that strip forests and take years to recover from. The tailings from those mines are toxic. While some of the newer production methods eschew the open-pit mines and instead process the sands underground or in situ, which is much cleaner, they still require additional energy to turn oil sands into usable crude. As a result, a barrel of oil-sand crude usually has a 10% to 15% larger carbon footprint than conventional crude over its lifetime, from the well to the wheels of a car. Given the massive size of the oil-sand reserve—nearly 200 billion recoverable barrels—that's potentially a lot of carbon. It's not surprising that environmentalists have loudly opposed the Keystone XL pipeline, which would send 800,000 barrels of oil-sand crude a day to the U.S. "There's enough carbon there to create a totally different planet," says James Hansen, a NASA climatologist and activist.

Tight-oil production isn't as polluting as extracting from oil sands, but it does make use of fracking, which has quickly become the most controversial technique in energy. Fracking fluids contain small amounts of toxic chemicals, and there have been allegations in Pennsylvania—where fracking has been used to produce shale natural gas—that it contaminates groundwater. The federal government is considering stricter regulations on the practice. "The federal rules have loopholes, and the state rules are too weak," says Amy Mall, a senior policy analyst for the Natural Resources Defense Council. "There are risks to groundwater, and there are risks to air." So far, there have been few complaints of water pollution from tight-oil wells in North Dakota and Texas, though

AARON KUO-DEEMER /THE NEW YORK TIMES/ REDUX

Oil consumption is growing rapidly in developing nations like China, putting pressure on prices and supplies. In the capital Beijing, a new wave of cars has resulted in severe traffic and choking air pollution.

both those states have notably oil-industry–friendly attitudes.

If tight-oil production spreads to more densely populated states like Ohio and California, both of which have shale plays, we could see those states gripped by the same controversies that have come with shale gas in Pennsylvania and New York. Sparse North Dakota is struggling to deal with the sudden influx of workers and equipment as well as the air pollution that results from tight-oil production. Even the oil industry is realizing that it needs to assuage public concerns. "We cannot ignore parts of the public that don't trust our industry and our ability to operate safely," Statoil CEO Helge Lund said at a 2011 energy conference. "This is a fundamental issue affecting us all."

The offshore drilling in Brazil's pre-salt reservoirs and in the Arctic waters being opened up by climate change is cleaner, but as seen with the Deepwater Horizon spill, if something goes wrong, it means catastrophe. If you think cleaning up an oil spill in the Gulf of Mexico was tough, try doing it in the remote, forbidding Arctic. But even greater than the immediate environmental danger posed by unconventional oil is the larger risk to the climate. One of the expected consolations of peak oil was the assumption that running out of conventional crude would finally force us to develop carbon-free alternatives such as wind and solar. Extreme oil means there will still be enough—more than 1 trillion barrels by one estimate—to keep cooking the planet if we decide to burn it all. Deborah Gordon, an expert at the Carnegie Endowment for International Peace, says that "21st-century oil is not 20th-century oil. New, unconventional oils are going to recarbonize global petroleum supplies."

So this is the future of oil: as costly as it is polluting. But if we can't ensure cheap oil, we can become more resilient when fuel becomes expensive. That requires continued improvements in energy efficiency. The U.S. has made some strides recently in that area (new vehicles get better mileage now than ever before), but it still lags behind the rest of the world. Obama's push to increase corporate average fuel-efficiency standards for vehicles to 55 miles per gallon by 2025 is vital. After all, doubling the mileage of your car is the equivalent of cutting the price of gasoline in half. Other kinds of energy alternatives must be developed to break the monopoly of crude, for environmental and economic reasons. Diversifying your energy supply is as wise as diversifying your portfolio. "We've got to develop every source of American energy, not just oil and gas but wind power and solar power, nuclear power and biofuels," Obama said in a speech in March 2011. "That's the only solution to this challenge."

From Brazil to Bismarck, human ingenuity (and tens of billions of dollars in investment) has extended the age of oil, as well as our anxiety about it. There's no reason the same formula can't eventually bring it to an end—on our terms.

The War On Coal

Activists cite public-health hazards in a new campaign against coal. Opponents say cleaner is too costly.

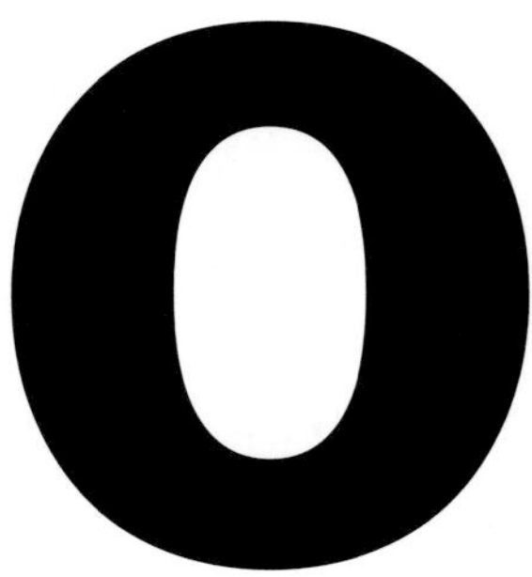

ON A 99°F JULY SUNDAY, THERE'S no cooler place to be in Chicago's Pilsen neighborhood than the public pool in Dvorak Park, where you can catch a fleeting breeze in this working-class, heavily Latino community. Unfortunately, the air in Pilsen isn't very cool—and it isn't very clean. Chicago's air on July 17, 2011, was so polluted that the government recommended that children and people with respiratory ailments—too common in a city that has nearly double the national asthma-hospitalization rate—limit their time outdoors. "People are getting sick in Chicago because of the air," says Brian Urbaszewski, director of environmental-health programs at the Respiratory Health Association of Metropolitan Chicago. "And it's people who are living in neighborhoods like Pilsen that are getting the worst of it."

That's due in part to the 450-foot brick smokestack that looms over Dvorak Park—the one the kids call "the cloudmaker." It belongs to the Fisk Generating Station, a 326-MW station just a couple of blocks from the park that's one of the oldest coal-fired power plants in the country. A 2010 report by the Clean Air Task Force (CATF), an NGO that focuses on air pollution, estimated that toxic emissions from the Fisk plant alone are responsible for 15 premature

BLACK RIVER PRODUCTIONS, LTD./MITCH EPSTEIN, COURTESY OF SIKKEMA JENKINS & CO., NEW YORK

In West Virginia, the coal industry is a part of everyday life—whether residents want it or not.

Even as coal dwindles in the U.S., miners like these seen in China are still producing vast amounts of it—and risking their health.

MICHAEL REYNOLDS—EPA/CORBIS

deaths a year and 200 hospitalizations. That's partly a result of Fisk's age—a grandfather clause in the Clean Air Act exempts older power plants in the U.S. from meeting some tougher regulations—but it also has to do with the fact that more people live within a mile of Fisk than any other coal plant in the country. Schools and playgrounds sit within sight of the smokestack. "You can feel it in your lungs when you live here," says Leila Mendez, a longtime Pilsen resident. "My hope is that it will just be closed."

Mendez and other activists in Chicago got their wish—in 2012 the corporate owner of the plants announced that they would be closed early. But Fisk won't be the first old coal plant to shut down because of pollution concerns—or the last. The powerful coal industry—which provides nearly half the electricity used by Americans, along with 30% of U.S. carbon emissions and a smoggy chunk of the nation's air pollution—is being attacked by an insurgency of environmentalists, regulators, and health advocates. In the wake of failed carbon cap-and-trade legislation and fizzling international climate talks, environmentalists concerned about global warming are taking on the coal industry from a different angle: public health. The Environmental Protection Agency (EPA) has developed a series of long-delayed regulations under the Clean Air Act that, if pursued aggressively, could make life very difficult for Big Coal. In the summer of 2011 Michael Bloomberg announced a $50 million gift to the Sierra Club's $150 million Beyond Coal campaign through his charitable foundation, marking coal pollution as the target of the New York City mayor's latest crusade. Bloomberg wants to turn coal into the new tobacco, to make it politically and culturally unacceptable because of the damage it does to everyone's health. "This is a public-health issue, just like our efforts to stop smoking or help with malaria," Bloomberg told TIME. "The pollutants and toxins are a big problem."

The campaign comes at a time when the world is becoming more coal-intensive. Global coal consumption grew by 7.6% in 2010; 30% of the world's energy now comes from coal, up from 25% five years ago. The rise is driven by emerging economies like China, the world's biggest coal consumer, which nearly tripled its coal use over the past decade. That hasn't helped coal-fighting efforts in the West, where stalling economies are struggling to keep up with gangbuster growth in China and other emerging markets. Coal exports from the U.S., the world's second largest coal producer and consumer, to Asian countries more than tripled last year, a boon for the industry. U.S. dependency on coal-fired electricity continues at home, which could result in higher energy costs and job losses if air-quality regulations are tightened. King Coal—still the cheapest source of electricity—is entrenched in the global energy system.

The Sierra Club, one of the country's oldest and largest green groups, with 1.4 million members, has already found early success fighting the growth of coal on a shoestring budget. So far, the club says, its Beyond Coal campaign has helped block more than 150 proposed coal plants across the country, using legal action and local opinion. With Bloomberg's help, it's stepping up the battle, seeking to shut existing coal plants. And the government is looking to crack down on air pollution as well. "No community should bear the burden of another community's polluters or be powerless to act against air pollution that leads to asthma, heart attacks, and other harmful illnesses," says EPA Administrator Lisa Jackson.

Pick Your Poison

"Is there a war on coal?" asks American Electric Power (AEP) CEO Michael Morris. "I think that's fair to say." Morris's opinion matters. AEP is one of the largest and most coal-dependent utilities in the country, and the company has not suffered the coming EPA regulations quietly. Morris says the rapid pace of the rules—utilities will have only three years to meet the tighter emissions standard on cross-border pollution—will cost the company billions of dollars and hundreds of jobs.

According to a study released in June 2011 by the National Economic Research Associates, an economics-consulting firm, the EPA's cross-border-pollution rule and its proposed mercury and toxics regulations will cost the industry $18 billion a year, create job losses, and increase the average American electricity bill by 11.5% by 2020. While those numbers are much higher than the EPA's estimates—and it's worth noting that the study was commissioned by the coal sector—any regulations that seriously take on coal power in the U.S. will have at least a short-term economic cost. And that cost will be especially heavy in places like Kentucky, where coal provides more than 90% of the state's electricity and some 18,000 mining jobs. But if states issue rules that raise the cost of coal power, Midwestern utility executives will start muttering darkly about power outages, while Republicans and Democrats alike in coal states will get angry. "Coal not only built this country, but it also built the skyscrapers of New York City," Democratic Senator Joe Manchin of West Virginia said in response to Bloomberg's pledge. "Without coal, the lights of that city would be dark and its economy would be devastated."

Healthy Air

But air-pollution regulations don't come alone with costs. They also deliver economic benefits. The EPA argues that the new cross-state-border rule will provide $280 billion in public-health benefits—fewer deaths, hospital stays, and sick days—at the cost of roughly $2.5 billion a year in plant upgrades. (It helps that the low price of cleaner-burning natural gas—thanks chiefly to the recent boom in shale-gas production—has reduced the cost of switching away from coal.) Another study, by researchers at the University of Massachusetts, estimated that the new regulations would actually lead to a net increase in jobs as utilities hired workers to overhaul their most antiquated plants. Indeed, a report in 2011 by the White House Office of Management and Budget found that EPA regulations have historically provided $4 in health and environmental benefits for every $1 they cost.

Coal-industry executives argue that they've been reducing air pollution over the years—and they have. The skyline is significantly cleaner than it was in the 1960s and '70s. But the smoke hasn't cleared, and as scientists look more closely at air pollution, they're finding danger at lower and lower levels. Invisible particles—bits of soot less than nine ten-thousandths of an inch wide—can penetrate the lungs and trigger inflammation, which can contribute to cardiovascular disease. Mercury, which can cause neurological damage in children, is present in trace amounts in some kinds of coal and can be released into the air when the coal is burned. (Mercury emissions increased more than 8% from 1999 to 2005 even as levels of other pollutants fell.) "For someone who is predisposed to wheeze, air pollution is likely to tip them over and make them wheeze," says Dr. Jerome Paulson, chair of the American Academy of Pediatrics' council on environmental health.

It makes Bloomberg worry as well, which is why his foundation is donating millions of dollars to the Sierra Club's Beyond Coal campaign. As the mayor of a city that struggles to meet air-pollution targets—in part because of haze from power plants hundreds of miles away—Bloomberg knows the toll of coal. The air pollution from coal is a threat to urban public health, one that Bloomberg, who has taken cigarettes out of the hands of angry smokers in New York City, is ready to fight. "Coal kills every day," he says. "It's a dirty fuel."

Even with an injection of Bloomberg bucks, the Sierra Club is still a long way from getting the U.S. to truly move beyond coal. Even though U.S. coal generation is lower now than it has been in years, coal is still far and away the single biggest source of electricity in the country. And then there are carbon emissions. While existing technology can reduce traditional pollutants like sulfur dioxide and particulates from coal combustion, there's still no commercially viable way to take the carbon out of coal. Pilot projects in the U.S. to build such "clean coal" plants have stalled largely because Congress has failed to enact limits on carbon emissions. "The U.S. simply isn't taking the steps needed to clean coal," says John Thompson, director of the coal transition project at CATF.

Given the enormous size of the challenge before them, environmentalists are going to need chutzpah as much as they do a checkbook. "We will devote more resources to moving America beyond coal than anything the Sierra Club has done in its 125-year history," says Michael Brune, the Sierra Club's executive director. "We will create a breakthrough." The war on coal—and there is one—is just beginning.

Nature at Risk

From polar bears in the Arctic to bluefin tuna in the sea, wildlife is threatened by climate change and the expanding human population. Conservationists are working hard to preserve nature, but the question remains: Is there room for all God's creatures?

MICHAEL DUVA—GETTY IMAGES

The sifaka lemurs are one of the more threatened species of primates in Madagascar.

NATURE AT RISK

The New Age Of Extinction

As the globe warms, more than the climate is endangered. Species are vanishing at a scary rate. We're the cause—but we're also the solution.

THERE ARE AT LEAST 8 MILLION UNIQUE SPEcies of life on the planet, if not far more, and you could be forgiven for believing that all of them can be found in Andasibe. Walking through this rain forest in Madagascar is like stepping into the library of life. Sunlight seeps through the silky fringes of the Ravenea louvelii, an endangered palm found, like so much else on this African island, nowhere else. Leaf-tailed geckos cling to the trees, cloaked in green. A fat Parson's chameleon lies lazily on a branch, beady eyes scanning for dinner. But the animal I most hoped to find, I don't see at first; I hear it, though—a sustained groan that electrifies the forest quiet. My Malagasy guide, Marie Razafindrasolo, finds the source of the sound perched on a branch. It is the black-and-white indri, largest of the lemurs—a type of small primate found only in Madagascar. The cry is known as a spacing call, a warning to other indris to keep their distance, to prevent competition for food. But there's not much risk of interlopers. The species—like many other lemurs, like

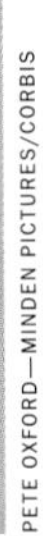
PETE OXFORD—MINDEN PICTURES/CORBIS

many other animals in Madagascar, like so much of life on Earth—is endangered and dwindling fast.

Madagascar—which separated from India 80 million to 100 million years ago before eventually settling off the southeastern coast of Africa—is in many ways an Earth apart. All that time in geographic isolation made Madagascar a Darwinian playground, its animals and plants evolving into forms utterly original. They include species as strange-looking as the pygmy mouse lemur—a chirping, palm-size mammal that may be the smallest primate on the planet—and as haunting as the carnivorous fossa, a catlike animal about 30 inches long. Some 90% of the island's plants and about 70% of its animals are endemic, meaning that they are found only in Madagascar. But what makes life on the island unique also makes it uniquely vulnerable. "If we lose these animals on Madagascar, they're gone forever," says Russell Mittermeier, president of the wildlife group Conservation International (CI).

That loss seems likelier than ever because the animals are under threat as never before. Once lushly forested, Madagascar has seen more than 80% of its original vegetation cut down or burned since humans arrived at least 1,500 years ago, fragmenting habitats and leaving animals effectively homeless. Unchecked hunting wiped out a number of large species, and today mining, logging, and energy exploration threaten those that remain. "You have an area the size of New Jersey in Madagascar that is still under forest, and all this incredible diversity is crammed into it," says Mittermeier, an American who has been traveling to the country for more than 25 years. "We're very concerned."

Madagascar is a conservation hot spot—a term for a region that is very biodiverse and particularly threatened—and while that makes the island special, it is hardly alone. Conservationists estimate that extinctions worldwide are occurring at a pace that is up to 1,000 times as great as history's background rate before human beings began proliferating. Worse, that die-off could be accelerating.

Price of Extinction

There have been five extinction waves in the planet's history—including the Permian extinction 250 million years ago, when an estimated 70% of all terrestrial animals and 96% of all marine creatures vanished, and, most recently, the Cretaceous event 65 million years ago, which ended the reign of the dinosaurs. Though scientists have directly assessed the viability of fewer than 3% of the world's described species, the sample polling of animal populations so far suggests that we may have entered what will be the planet's sixth great extinction wave. And this time the cause isn't an errant asteroid or megavolcanoes. It's us.

Through our growing numbers, our thirst for natural resources, and, most of all, climate change—which,

Madagascar's Vanishing Treasures

Madagascar is truly like no place on Earth. The island separated from India 80 million to 100 million years ago, which gave its native wildlife plenty of time to evolve in isolation. As a result, Madagascar has more endemic species for its size than any other country on Earth. But that wildlife is under extreme pressure—and if they're gone in Madagascar, they're gone for good. Some of the endangered species under threat:

1) Aye-Aye: A lemur found in Madagascar trees. It has a long middle finger it uses to dig insects out of the bark. **2) Ploughshare tortoise:** A critically endangered animal. **3) Golden mantella:** These frogs grow only 1 inch long. **4) Panther chameleon:** Beautiful, but extremely territorial lizards. **5) Conocephaline katydid:** A glorious pink insect. **6) Fossa:** These cat-like animals are Madagascar's largest mammalian predator. **7) Comet moth:** A rain-forest native insect. **8) Pygmy mouse lemur:** At only 5 inches long, this is one of the smallest primates in the world.

CLOCKWISE FROM TOP LEFT: MARK CARWARDINE—PETER ARNOLD/GETTY IMAGES; KEVIN SCHAFER—GETTY IMAGES; CHRIS HELLIER—CORBIS; PETE OXFORD—MINDEN PICTURES; THOMAS MARENT—MINDEN PICTURES; PAUL BERTNER—GETTY IMAGES; ALBERT LLEAL—MINDEN PICTURES; CAROL POLICH—GETTY IMAGES

6
4
7
5
8

The gorgeous baobab or upside-down trees of Madagascar look like something out of a Dr. Seuss story.

FROM LEFT: THOMAS MARENT—MINDEN PICTURES; MICHAEL FAY—NATIONAL GEOGRAPHIC/GETTY IMAGES

by one reckoning, could help carry off 20% to 30% of all species before the end of the century—we're shaping an earth that will be biologically impoverished. A 2008 assessment by the International Union for Conservation of Nature (IUCN) found that nearly 1 in 4 mammals worldwide was at risk for extinction, including endangered species like the famous Tasmanian devil. Overfishing and acidification of the oceans are threatening marine species as diverse as the bluefin tuna and reef-forming corals. "Just about everything is going down," says Simon Stuart, head of the IUCN's species-survival commission. "And when I think about the impact of climate change, it really scares me."

Scary for conservationists, yes, but the question arises, Why should it matter to the rest of us? After all, nearly all the species that were ever alive in the past are gone today. Evolution demands extinction. When we're using the term extinction to talk about the fate of the U.S. auto industry, does it really matter if we lose species like the Holdridge's toad, the Yangtze River dolphin, and the golden toad, all of which have effectively disappeared in recent years? What does the loss of a few species among millions matter?

For one thing, we're animals too, dependent on this planet like every other form of life. The more species living in an ecosystem, the healthier and more productive it is, which matters for us—a recent study by the World Wildlife Fund (WWF) estimates the economic value of the Amazon rain forest's ecosystem services to be up to $100 per hectare (about 2½ acres). When we pollute and deforest and make a mess of the ecological web, we're taking out mortgages on the Earth that we can't pay back—and those loans will come due. Then there are the undiscovered organisms and animals that could serve as the basis of needed medicines—as the original ingredients of aspirin were derived from the herb meadowsweet—unless we unwittingly destroy them first. "We have plenty of stories about how the loss of biodiversity creates problems for people," says Carter Roberts, WWF's president.

Forests razed can grow back, polluted air and water can be cleaned—but extinction is forever. And we're not talking about losing just a few species. In fact, conservationists quietly acknowledge that we've entered an age of triage, when we might have to decide which species can truly be saved. The worst-case scenarios of habitat loss and climate change—and that's the pathway we seem to be on—show the planet losing hundreds of thousands to millions of species, many of which we haven't even discovered yet. The result could be a virtual genocide of much of the animal world and an irreversible impoverishment of our planet. Humans would survive, but we would have doomed ourselves to what naturalist E.O. Wilson calls the Eremozoic Era—the Age of Loneliness.

A 2004 study estimated that global warming could drive a million species to extinction by mid-century.

So if you care about tigers and tamarins, rhinos and orangutans, if you believe Earth is more than just a home for 7 billion human beings and counting, then you should be scared. But fear shouldn't leave us paralyzed. Environmental groups worldwide are responding with new methods to hinder new threats to wildlife. In hot spots like Madagascar and Brazil, conservationists are working with locals on the ground, ensuring that the protection of endangered species is tied to the welfare of the people who live closest to them. A strategy known as avoided deforestation goes further, incentivizing environmental protection by putting a price on the carbon locked in rain forests and allowing countries to trade credits in an international market, provided that the carbon stays in the trees and is not cut or burned. And as global warming forces animals to migrate in order to escape changing climates, conservationists are looking to create protected corridors that would give the species room to roam. It's uncertain that any of this will stop the sixth extinction wave, let alone preserve the biodiversity we still enjoy, but we have no choice but to try. "We have a window of opportunity," says Kassie Siegel, director of the Climate Law Institute of the Center for Biological Diversity (CBD). "But it's slamming shut."

To Save the Species, Save the People

Madagascar, which Mittermeier calls the "hottest of the hot spots," is where all the new strategies can be road-tested. In 2003, after decades when conservation was barely on the government's agenda, then-President Marc Ravalomanana announced that the government would triple Madagascar's protected areas over the following five years. That decision helped underfunded parks like Andasibe's, which protects some of the last untouched forest on the island. "You can't save a species without saving the habitat where it lives," says WWF's Roberts.

Do that right, and you can even turn a profit in the process. In Madagascar, half the revenues from national parks are meant to go to the surrounding communities. The reserves in turn help sustain an industry for local guides like Razafindrasolo. In a country as poor as Madagascar—where 61% of the people live on less than $1 a day—it makes sense to give locals an economic stake in preserving wildlife rather than destroying it. "If you don't get the support of the people living near a conservation area, it's just a matter of time before you'll lose [the area]," says Steven Sanderson, president of the Wildlife Conservation Society (WCS).

Well-run ecotourism can provide support for conservation, but even the best parks might be hard-pressed to compete with the potential revenues from logging,

Slash-and-burn agriculture has devastated much of Madagascar.

poaching, or mining. The strategy of avoided deforestation, however, offers much more. Rain forests like those in Madagascar contain billions and billions of tons of carbon; destroying the trees and releasing the carbon not only kills local species but also speeds global warming. Proposals in the global climate negotiations would allow countries to offset some of their greenhouse-gas emissions by paying rain-forest nations to preserve their trees. It's win-win, with both the climate and the critters getting a boost. In eastern Madagascar, CI and WCS are working together to protect about 865,000 acres in the Makira Forest with a range of carbon investors that include Mitsubishi and Pearl Jam. Closer to Andasibe, CI and its partners are hiring villagers to plant trees on eroded land, which creates corridors to connect fragmented habitats, may earn carbon revenues, and provides needed employment. "We're bringing back the shelter of the forests, and we don't have to cut trees," says Herve Tahirimalala, a Malagasy who is paid $100 a month to work on the project.

The corridors created by CI's Andasibe tree-planting program show how a small tweak can reduce the

Animals Under Siege

The earth supports amazing biodiversity, but much wildlife is coming under threat from one particularly successful species: us.

Human Encroachment

With a population of over 7 billion—and set to pass 9 billion by 2050—human beings are crowding out other species through sheer numbers. As a rising middle class in countries like India begins consuming at Western rates—including both food and energy—we could leave little room for wildlife.

Illegal Wildlife Trade

The buying and selling of endangered live animals—as pets, performers, and more—is a new and growing menace to wildlife. Total wildlife trade is believed to be the second largest direct threat to many species, after habitat loss. The problem has become so severe that conservationists have coined the term empty-forest syndrome to describe habitats that remain standing but have been stripped of the animals that once lived there. Southeast Asia has long been a hub of the wildlife trade; almost anything can be found in the live markets of Bangkok or Guangzhou. And the Internet has allowed the sale of species to go global.

Climate Change

Even if we do manage to reduce forest loss and stop wildlife trade, a greater threat looms on the horizon: global warming. As the climate changes rapidly, the territory to which species have adapted may become unlivable faster than they can respond. The Intergovernmental Panel on Climate Change has reported that warming could put as much as 70% of species at a greater risk of extinction—with polar animals like the king penguin especially vulnerable.

Poaching

From the illegal slaughter of elephants for their ivory to the hunting of apes and gorillas for food, poaching remains a serious threat to some of the world's most charismatic endangered species. The rising global trade in animal parts—like tiger penises, prized as an aphrodisiac in China—has made killing even more profitable. In the Congo Basin, three-fifths of all large mammals are being hunted at unsustainable rates—many for "bushmeat," which is increasingly popular in African markets.

Deforestation and Habitat Loss

Tropical forests are home to the greatest concentration of biodiversity on the planet—but when the trees are lost, species soon follow. From 1990 to 2005, the world lost 172 million acres (70 million hectares) of forest—much of it in South American countries like Brazil, where deforestation has accelerated as land is cleared for pastures.

CLOCKWISE FROM TOP RIGHT: FRANS LANTING—CORBIS; MICHAEL NICHOLS—NATIONAL GEOGRAPHIC STOCK; APICHART WEERAWONG—AP; JOSH WESTRICH—CORBIS; PAWEL KOPCZNSKI—REUTERS

species-killing effects of climate change—but also how longer-term fixes are needed. Fragmented habitats are problematic because many endangered species wind up trapped in green oases surrounded by degraded land. As global warming changes the climate, species will try to migrate, often right into the path of development and extinction. What good is a nature reserve—fought for, paid for, and protected—if global warming renders it unlivable for the animals it's meant to shelter? "Climate change could undermine the conservation work of whole generations," says Larry Schweiger, president of the National Wildlife Federation. "It turns out you can't save species without saving the sky."

That will mean reducing carbon emissions as fast as possible. In the U.S., the CBD has made an art out of using the Endangered Species Act, which mandates that the government prevent the extinction of listed species, to force Washington to act on global warming. The CBD's Siegel led a successful campaign to get the Bush Administration to list the polar bear as threatened by climate change, and she expects more species to follow. "Polar bears are the canaries in the coal mine," says Siegel.

Why We Can't Wait

What's especially frightening is how vulnerable even the best conservation work can be to rapid changes—both climatic and governmental. Over the past few years, Madagascar has fallen into a political abyss, with Andry Rajoelina—the former mayor of Antananarivo, the capital—forcing former President Ravalomanana from office on the heels of deadly protests. As a result, development aid to the desperately poor country has been delayed, and conservation work has been disrupted. Reports have filtered back of armed gangs stepping into the vacuum to illegally log the nation's few remaining forests. "They're ripping out valuable timber as quickly as they can," says Mittermeier.

News like that can tempt even the staunchest defenders of wildlife to simply surrender. And why shouldn't they? In a world where hundreds of millions of human beings still go hungry and the global recession has left all but the wealthiest fearing for their future, it's easy to wonder why we should be concerned about the dwindling of the planet's biodiversity.

The answer is that we can't afford not to. The same natural qualities that sustain wildlife—clean water, untainted land, unbroken forests—ultimately sustain us as well, whether we live in a green jungle or a concrete one. But there is an innate value to untrammeled biodiversity too—one that goes beyond our own survival. When that is lost, we are irretrievably diminished. "We live on a very special planet—the only planet that we know has life," says Mittermeier. "For me, conservation is ultimately a moral obligation and simply the right thing to do." That leaves us a choice. We can save life on this special planet, or be its unwitting executioner.

SHAUL SCHWARZ

The golden sargassum makes the Sargasso Sea the nursery of the oceans—which is why Sylvia Earle wants to save them.

Code Blue

Scientists want to save our oceans the simple way: Protect them. But organizing marine-protected areas for the seas will be harder than it sounds. Do we care enough?

THE SARGASSO SEA HAS NO shores. THE 2-MILLION-square-mile body of water in the middle of the Atlantic is defined by two features: the ocean currents forming the North Atlantic subtropical gyre, which cycles around the sea, and sargassum, the free-floating golden-brown seaweed. The sargassum can be found scattered throughout the sea, sometimes entwined in vast waterborne mats. When Christopher Columbus encountered the sargassum while crossing the Atlantic, he ordered his men to fathom the depths, believing he had struck land. The oceanographer Sylvia Earle, though she prefers not to think in land-based metaphors, calls the sargassum "the golden rain forest of the sea," a base for scores of juvenile creatures, a floating nursery in a sea that was long believed to be a watery desert. She has traveled to Bermuda, on the western fringes of the Sargasso Sea, to see the sargassum and the ocean life she has worked for decades to protect. "The sargassum is the shelter," she says as her boat passes beyond Bermuda's coral reefs. "It is the island in the stream."

It's an island that can be tough to find. The late-summer hurricanes that plague the Atlantic had churned the waters off Bermuda like an eggbeater, breaking up the biggest sargassum mats. But eventually, after the shallows give way to the cobalt blue of the open ocean, the signal is given: Sargassum, ho! Woven into the weed are pink filaments of coral spawn, eggs, and sperm together—"a starter kit for a new coral reef," according to Earle, who slips over the side and into the ocean. The warm water is deep, but the sun can shine through the sea more than 200 feet down, the

It's not just fish that depend on the oceans and the sea. From top left, the European eels spawn in the Sargasso Sea before spreading out into the world; polar bears are marine mammals that depend on sea ice as their habitat; the staghorn coral help make up Australia's Great Barrier Reef; the Hawksbill sea turtle is on the brink of extinction.

light broken only by the golden shade of sargassum mats. Beneath the surface are thousands, perhaps millions of coral eggs, each smaller than the period at the end of this sentence, leaving Earle to swim through a sea of stars. After nearly an hour, she emerges from the water, dripping and smiling. "Every time I see a big batch of eggs, I know another generation is coming," she says. "I know there's hope."

In her more than 50-year career in ocean science, Earle has studied the deep, dived it, and written about it. Now she's trying to save the oceans, and she's focusing on one simple idea: Protect them. Create true reserves—marine protected areas (MPAs)—on the high seas, sanctuaries for everything that swims, just as governments have created parks on land. About 12% of the planet's land surface is conserved in some way, but little more than 1% of the 139.5 million square miles of the oceans have any protection whatsoever.

To Earle, that discrepancy helps explain why we've treated the oceans as a "supermarket and a sewer." First, through overfishing: Factory trawlers and long-liners, operating with generous government subsidies, have stripped the seas of their abundance, and many scientists estimate that we've lost at least 90% of large predatory fish like sharks and swordfish as populations of once abundant commercial species like bluefin tuna have crashed. Our appetite for fish has disrupted the aquatic food chain, degrading the coral reefs that sustain ocean life around the planet. Pollution and fertilizer runoff from agriculture have helped push one-third of the more than 700 reef-building corals close to extinction; 70% of all coral reefs could be gone by mid-century.

Even worse are long-term changes to the very chemistry of the seas. The oceans have gradually warmed, depriving species of nutrients and triggering deadly coral-bleaching events. Unusually high temperatures this year could wipe out coral around the planet, accelerating the destruction of the most valuable habitats in the oceans. Thanks to all the fossil fuels humans have burned, the seas have also become more acidic because dissolving CO_2 in water lowers the pH—the oceans store 50 times as much carbon as the atmosphere does—and that change will have consequences no one can predict, though none are likely to be positive. The seas seem as invulnerable as they are immense. But if we thought the oceans were too vast for human beings to affect or we were counting on the waters to purify themselves, we were wrong. "The loss we could suffer goes beyond aesthetics," says Josh Reichert, managing director of the Pew Environment Group. "It's a loss to ourselves."

Earle aims to stop it. She has launched the nonprofit Mission Blue to build support for marine sanctuaries around sensitive hot spots—she calls them hope spots—in the endangered oceans, including the high seas beyond any one country's control. If her goal is audacious—we don't even have the legal institutions yet to protect international waters—so is Earle. "I can't think of many

FROM LEFT: WILL MEINDERTS—FOTO NATURA/MINDEN; JOSEF FRIEDHUBER—VETTA/GETTY IMAGES; FRANS LANTING—CORBIS; JAMES R.D. SCOTT—GETTY IMAGES

others who've been as persistent and vocal and forward-thinking on the oceans as Sylvia," says Greg Stone, chief scientist for oceans at Conservation International. "The world is opening up to her message." The often fractious marine-conservation movement—along with new corporate allies like Google—is making a concerted push for attention under the Mission Blue banner, focusing on reducing overfishing and expanding protection. The island nations of the central Pacific—home to some of the last truly pristine open waters on the planet—are moving to create a network of protection across their oceanscape, and the Obama Administration has announced an oceans policy that involves managing U.S. waters, not just exploiting them. "This idea's time has come," says venture capitalist David Shaw, who has devoted time and money to Earle's ideas. "And Sylvia embodies it."

From Pioneer to Protector

When Earle began working in marine science in the 1950s, there were few women in the field. She was allowed on a scientific voyage in the early '60s only after she agreed to help with the dishes and the cooking. She began diving off Florida at 17—using a helmet, compressed air, and weights—and by 1970 she was an aquanaut, leading an all-female team of scientists who lived and worked for two weeks in an undersea station off a Caribbean island. It was then that she learned to love diving deep, seeing the ocean far below and far away from human beings.

The desire to go deeper eventually pushed Earle out of academia. She co-founded a company to design and build her own subs, and in 1979 she set the record for solo untethered diving, walking on the seafloor 1,250 feet below the surface in a JIM, or atmospheric diving suit. She's spent about 7,000 hours diving—that's nearly 10 months of her life, sleeping and waking—earning her the nickname Her Deepness. At 75, the elfin Earle still dives. "Every time I slip into the ocean," she says, "it's like going home."

She began her career in the heroic age of ocean exploration, when Jacques Cousteau roamed the seas in his ship *Calypso* and the *Trieste* went to the bottom of the Mariana Trench. But she's also seen the degradation of the oceans up close—the tortures endured by her home waters in the Gulf of Mexico, the steady evidence that we were fishing out the seas. And she is aware that we've lost far more than we can know. There is evidence, sketchy but still there, that sea life was once far more robust than it was even in our recent memory, that cod off New England were once so abundant that the fish jumped onto ships, and oysters so plentiful that 700 million were sold in New York City in 1880. The result is what the French marine scientist Daniel Pauly has called "shifting baselines": We can't tell how bad it's gotten because we don't remember how good it was. "The seas today are at very low abundance for many of the things that we rely on for food," says Callum Roberts, a marine-conservation biologist at England's University of York. "Many species have fallen a thousandfold from what they once were. This is new territory for all of us. But the Sargasso Sea could provide a road map for creating more protection elsewhere."

Earle has always been involved in conservation—she was the chief scientist at the National Oceanic and

Earle is know as "Her Deepness," both for her long experience with diving and her passion for saving the oceans.

SHAUL SCHWARZ—GETTY IMAGES

Atmospheric Administration in the early 1990s and is an explorer in residence at the National Geographic Society—but over the past few years, she's looked to intensify those efforts. Marine conservation, however, has always been a stepchild in environmentalism, and the groups involved are often small and fractured. Earle was looking for a big stage, and she got one in 2009, when she addressed Technology, Entertainment, Design (TED), the major conference that connects tech heavyweights with innovative scientists and thinkers. Every year TED grants a prize to one speaker who makes a "wish to change the world." Earle's was simple: "a campaign to ignite public support for a global network of marine-protected areas." She won the prize. "We recognized that her mission to save the ocean is one that can only be achieved if we harness all of the resources available to us," says Chris Anderson, TED's curator.

Parks of the Sea

MPAS aren't a new idea. Countries have been giving some protection to priceless pieces of the ocean, like Australia's Great Barrier Reef, for years. The MPAs have even been growing in number recently. In 2006, President George W. Bush created the Papahanaumokuakea Marine National Monument in northwestern Hawaii, protecting nearly 140,000 square miles of ocean, and in 2010 Britain created the largest MPA in the world, around the Chagos Archipelago in the Indian Ocean. Studies have shown that MPAs can allow fish and coral to recover from exploitation, and they have spillover benefits for unprotected waters outside their borders. But MPAs still make up just a tiny fraction of total ocean area, and in most of them, fishing is still allowed. (Imagine if hunting were permitted in Yellowstone Park and you have an idea of the difference between land conservation and its blue counterpart.) And there is virtually no protection for the high seas, the 64% of the oceans that is beyond the jurisdiction of any single nation. (Governments have claim to the waters in their exclusive economic zones, extending 200 miles beyond their coastlines.) Though there are a number of global bodies that deal with parts of the high seas, from regional fishery organizations to the International Seabed Authority, there's no clear way to create reserves in international waters. "The high seas are the least protected areas in the world," says Elliott Norse, head of the Marine Conservation Institute.

That's why Earle has decided to focus on creating MPAs in the high seas, the heart of the ocean. As part of her prize, she received support from TED to form Mission Blue and launched it with a special TED conference held on board a ship off the Galápagos Islands in April of 2010. (She chose the Galápagos for more than just the scenery. Although 97% of the islands' land is strictly protected, just 3% of the surrounding ocean is, and the difference shows.) Philanthropists, venture capitalists, and celebrities—including the greenish actors Edward Norton and Leonardo DiCaprio—listened to marine scientists and environmentalists detail the global ocean crisis; the result was $17 million in commitments for a number of initiatives, including the creation of the first real high-seas MPA, in the Sargasso. "This is new territory for all of us," says Dan Laffoley, marine vice chair for the International Union for Conservation of Nature's World Commission on Protected Areas. "The Sargasso Sea could provide a road map for creating more protection elsewhere."

Thanks to the work of scientists at the Bermuda Institute of Ocean Sciences (BIOS), one of the oldest oceanographic institutes in the world, parts of the Sargasso Sea have been more closely studied than almost anywhere else in the open ocean. And there's something magical about the Sargasso, a symbol of borderless mystery. The European and American eels hatch in the Sargasso Sea, then travel to coastal and inland waters, in one of the most enigmatic migrations in the natural world, before returning to the sargassum to spawn and die. And while its remoteness has so far saved the Sargasso Sea from the damage done to other parts of the ocean, there are emerging threats, including the risk that the sargassum might be harvested for biofuel or fertilizer, robbing this aquatic nursery of its cradles.

Earle, for her part, is an absolutist. She won't eat fish any longer—"I don't eat my dive buddies," she says—and urges others to do the same. When she talks about creating protected areas, she wants true marine sanctuaries, where fishing would be largely off limits, at least for a time. That brings opposition from the fishing industry, which has long been skeptical of the value of MPAs and doubtful that the seas are really in danger of being overfished. That's hard to believe. Fishermen have been able to mask the drop in fish populations only by going farther from shore and going farther down the aquatic food chain while taking advantage of improving subsidies and technology. Eventually they'll run out of room and fish. But MPAs alone, no matter how large, can't save vulnerable areas of the ocean from the impact of warming and, worse, gradual acidification, which will alter the chemistry of the oceans in a way the world hasn't seen for millions of years. "Once you screw with acidification, no one knows what's going to happen," says Tony Knap, former executive director of BIOS.

To Earle, however, climate change and its effects are just more reasons to get serious about carving out MPAs. Areas of the ocean with protection are more resilient to nearly all threats than those left undefended. And even fishermen should support the idea of marine reserves. Research from University of British Columbia economist Rashid Sumaila found that overfishing leads to global catch losses costing the industry up to $36 billion a year. If the global fish population is a bank account, we're cutting into our principal by overfishing, and we should all know by now that's not sustainable. Creating reserves and giving fish a breather allows the interest to build back up.

Bringing Order to the Open Seas

What this means, ultimately, is that we can no longer treat the open ocean as ungoverned space—for our sake (Earle is fond of reminding people that the oxygen in every other breath comes from the ocean) and the sake of all that lives beneath the surface. The good news is that there's evidence of a sea change in attitude. In 2010, the Obama Administration created a federal oceans policy that aims to govern U.S. waters in a comprehensive manner instead of keeping fishing, recreation, and energy in separate silos. The National Geographic Society has taken Earle's Mission Blue and built on it, seeking to establish a unified and global ocean-conservation movement. (Earle herself has launched a side group called SEAlliance, which will focus chiefly on MPAs.) Google is getting into the action. After Earle pointed out that the Google Earth application treated the seas as one big blue blank, the company added new functions that allow users to plunge into the deep. The nations of the Pacific, led by the tiny island of Kiribati, are working toward creating the Pacific Oceanscape, a linked, protected area in the middle of the world's biggest body of water. "We need to move beyond our coasts," says President Anote Tong of Kiribati. "The ocean is much larger than any one nation."

The ocean is bigger than all of us, and yet each of us can affect it via our actions. That might be the scariest news of all for the blue. If the oceans are suffering in a world with 7 billion human beings, more than 60% of whom live on or near a coastline, what will happen when there are 9 billion or more? "It's really hard to conjure up much of a sense of well-being, because it just keeps getting worse," says oceanographer Jeremy Jackson—aptly known in science circles as Dr. Doom.

Think about it too long and it's easy to sink into despair, but Earle floats above—or perhaps below—it. If it becomes too much, she can always slip into the sea, where she's spent so much of her life, moving with a balletic grace, in sync with the sea life she loves. On her most recent dive in the Sargasso Sea, Earle drifts past an algae-covered shipwreck, one of scores of vessels claimed by Bermuda's reefs. Seven thousand hours of diving, and every minute beneath the waves still excites her. She lands gently on the seafloor, following the path of a tiny arrow crab, one of the few signs of life in the gorgeous but empty waters. The sediment rises around her, and she disappears in a cloud of underwater dust. She's at home.

Vicious Cycles

The debate over whether Earth is warming up is over. Now we're learning that climate disruptions feed off one another in accelerating spirals of destruction. Scientists fear we may be approaching the point of no return.

THE GREENHOUSE EFFECT

Without the greenhouse effect, life on Earth would not be possible. Energy from the sun is absorbed by the planet and radiated back out as heat. Atmospheric gases like carbon dioxide trap that heat and keep it from leaking into space. That's what keeps us warm at night.

But as humans pour ever increasing amounts of greenhouse gases into the atmosphere, more of the sun's heat gets trapped, and the planet gets a fever

SUNLIGHT

HEAT

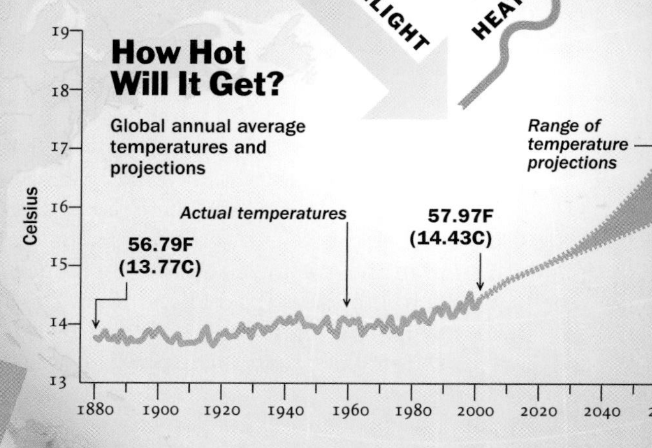

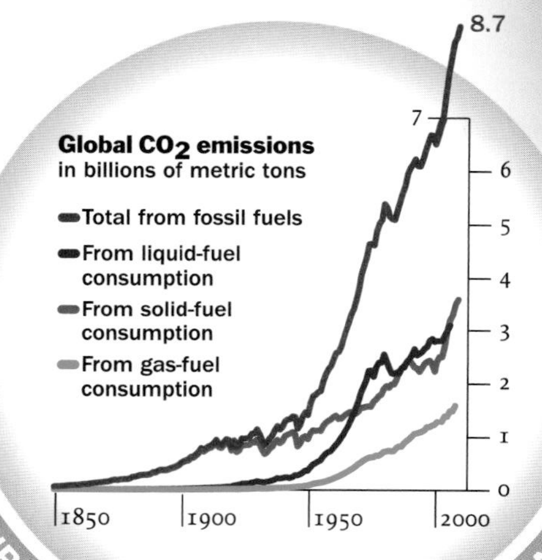

BURNING FOSSIL FUELS RELEASES CARBON

BURNING FORESTS

REDUCES OXYGEN AND INCREASES DROUGHT

MELT POLAR ICE AND PERMAFROST

Plants take in CO_2

Fires release carbon

Less carbon absorbed

Soil dries out

Sea-ice average 1979–2000

▲

FUELING THE FIRE The amount of carbon dioxide in the atmosphere is climbing fast. Most of it comes from burning fuels for energy—gasoline in cars or coal for electricity, for example. The U.S., with less than 5% of the world's population, produces nearly one-quarter of all greenhouse gases.

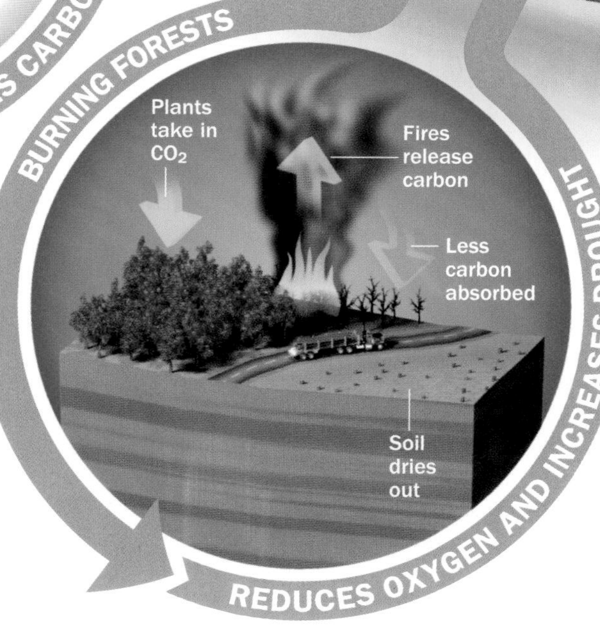

SPREADING THE PAIN ► Deforestation, through clear-cutting or burning, sows havoc far beyond the affected area. The fires release still more carbon into the atmosphere, fewer plants survive to convert CO2 into oxygen, and scorched soil absorbs more heat and retains less water, increasing droughts.

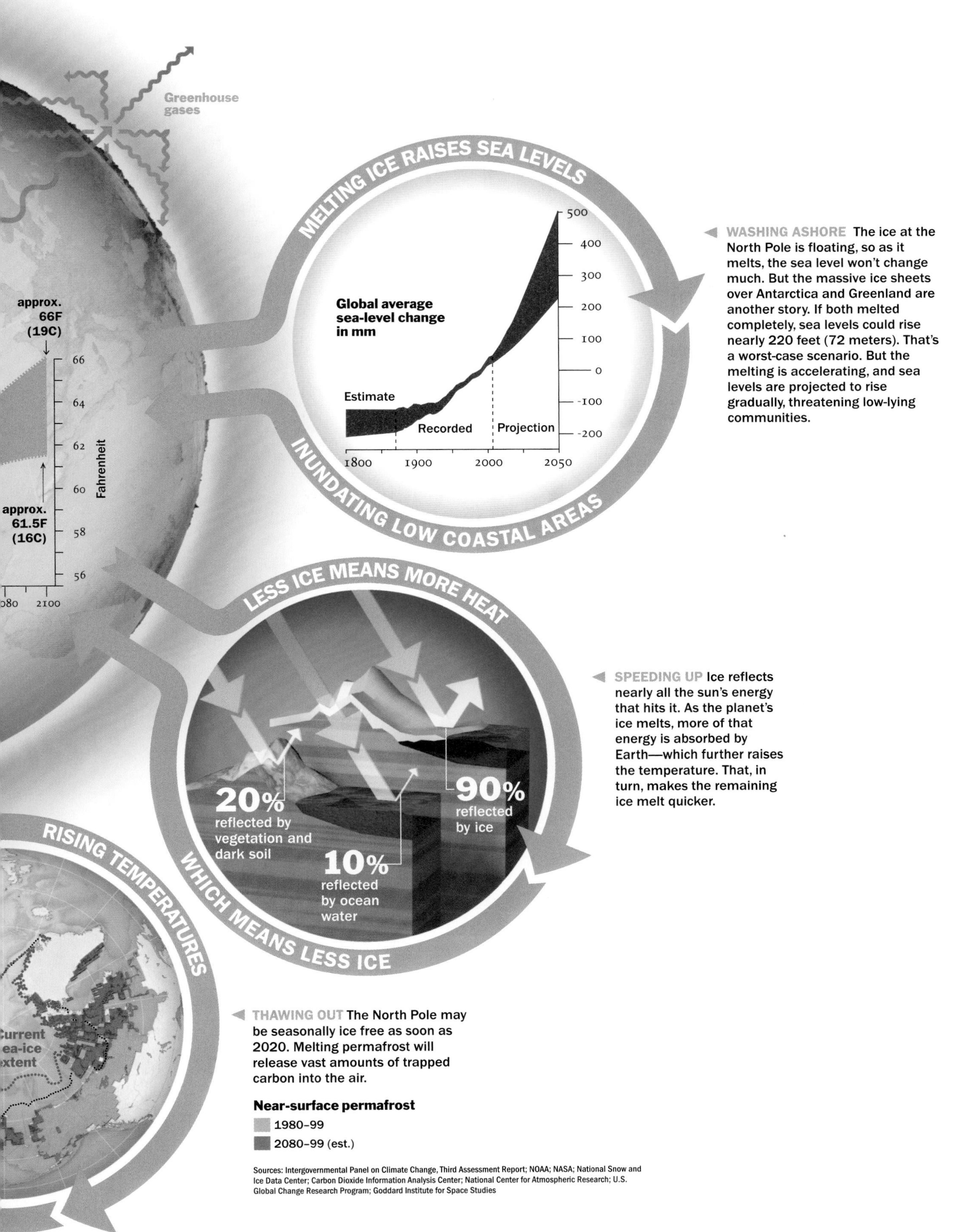

WASHING ASHORE The ice at the North Pole is floating, so as it melts, the sea level won't change much. But the massive ice sheets over Antarctica and Greenland are another story. If both melted completely, sea levels could rise nearly 220 feet (72 meters). That's a worst-case scenario. But the melting is accelerating, and sea levels are projected to rise gradually, threatening low-lying communities.

SPEEDING UP Ice reflects nearly all the sun's energy that hits it. As the planet's ice melts, more of that energy is absorbed by Earth—which further raises the temperature. That, in turn, makes the remaining ice melt quicker.

THAWING OUT The North Pole may be seasonally ice free as soon as 2020. Melting permafrost will release vast amounts of trapped carbon into the air.

Sources: Intergovernmental Panel on Climate Change, Third Assessment Report; NOAA; NASA; National Snow and Ice Data Center; Carbon Dioxide Information Analysis Center; National Center for Atmospheric Research; U.S. Global Change Research Program; Goddard Institute for Space Studies

The Urban Factor

Few people think of crowded cities as good for the environment, but the truth is that urban living is easier on the planet than settling in suburbs or rural areas. That's why cities like New York, Stockholm, and Abu Dhabi are leading the world in sustainability.

DAVID ARKY—CORBIS

Desert Dreams

Abu Dhabi is investing billions in oil and gas profits to turn itself into the world's leader in renewable energy.

THE DOORS SWISH SHUT AND with the press of a touch-screen button, the Personal Rapid Transit (PRT) car is off, gliding through the tunnels beneath Abu Dhabi's new Masdar City. The sleek four-passenger vehicle, which looks like something out of the movie *TRON: Legacy*, runs on an electric motor, making it clean and carbon-free. There are no tracks—the car is autonomous, driven by a computer that charts direction with the help of tiny magnets embedded in the road. When my PRT car senses another vehicle waiting in our parking space, it stops and waits for the area to clear, avoiding a collision. PRT is meant to be the future of mass transit within cities, with the environmental benefits of buses and trains but the freedom of a private vehicle. Yet as my car pulls into an open docking bay, I can't help thinking there's something slightly silly about all this. For all the technology, which isn't cheap, the PRT has taken me to its one and only stop, maybe half a mile (800 meters) from the starting point. For a lot less cost—and not much more time—I could have used a much older form of transport: my legs.

In a nutshell, that is what's good and bad about Masdar. Back in 2007, the government of Abu Dhabi, a Middle Eastern emirate that controls 8% of the world's oil reserves, announced that it would build "the world's first zero-carbon city," a custom-designed settlement called Masdar. (The word means source in Arabic.) It would rely entirely on renewable energy—mostly solar—and would produce zero waste. It would be home to a university dedicated to the study of sustainability, as well as attract the best companies in clean tech. There would be no traditional cars inside the city—all transportation was to be via PRT vehicle—and the city would use half the energy of a settlement of the same size. The urban layout by the green-minded British architect Norman Foster would combine classic Arab design with 21st-century technology. Masdar would be a living lab for a greener, cleaner future and a bridge for Abu Dhabi as it prepared for a day when its oil riches finally ran out. "We will position Abu Dhabi as the hub of future energy," Sultan Ahmed Al Jaber, Masdar's CEO, told me in January 2008.

Abu Dhabi's leadership was all the more necessary at a moment when once-vibrant green businesses were flagging, thanks in part to the financial crisis and plummeting price of oil at the time. In the U.S. and Europe,

The Masdar Institute, top, is the first phase of the larger city to be completed. Solar projects, below, will help power the city—key in the blazingly hot desert environment of Abu Dhabi.

FROM TOP: DUNCAN CHARD—*THE NEW YORK TIMES*/REDUX; PIERO OLIOSI—POLARIS

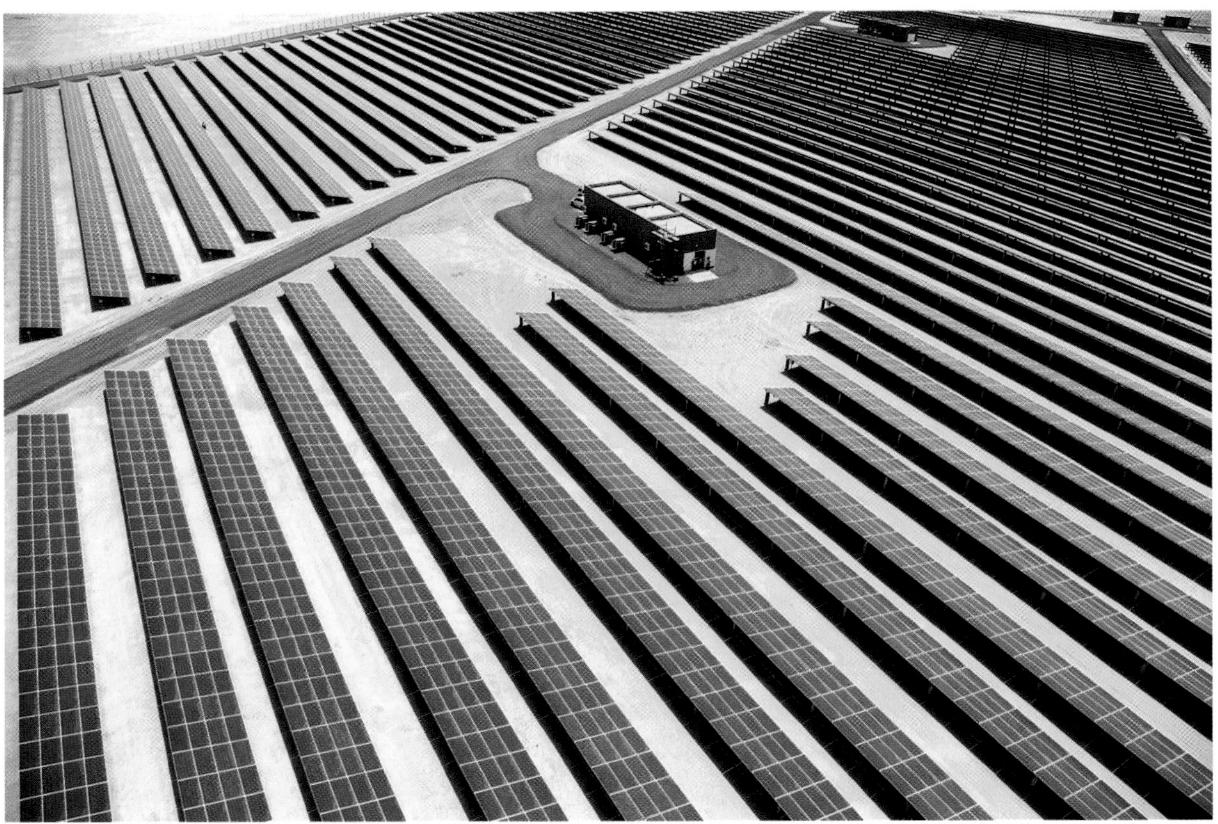

The Middle East is the ideal environment for solar power—though panels have to be cleaned regularly because of loose sand.

new wind- and solar-power installations were slowing and energy startups were laying off workers. But it was still full speed ahead in Abu Dhabi; when I visited the city again in 2009, it was hosting the World Future Energy Summit (WFES), which attracted more than 16,000 visitors and companies that ranged from General Motors to small Chinese solar manufacturers. At a moment when the Obama Administration in Washington was struggling to get its ambitious green agenda on track and international climate-change talks had ground to a disappointing halt, Abu Dhabi kept the momentum going at its summit by announcing that at least 7% of its electricity would come from renewable sources by 2020, up from virtually zero at the time.

Other plans included a thin-film solar factory, along with investments in wind and solar and in carbon-trading projects around the world. Most significantly, Masdar was pioneering model carbon-capture and sequestration projects with the energy and mining giants BP and Rio Tinto that would take CO_2 emissions from industries in the emirate and store the gas in abandoned oil wells. Since even the most optimistic energy experts predict we'll be burning fossil fuels for decades, perfecting carbon capture is vital to controlling emissions—and who would be better suited to cleaning up fossil fuels than an emirate that produces nearly 3 million barrels of oil a day? The desert country might have been more responsible than most for global warming, but it was doing more than its part to stop it. "We are looking beyond the current financial crisis," Al Jaber said in 2009. "But all our projects are still proceeding."

Fast-forward several years and the plans have changed. Masdar City was originally scheduled to be completed by 2015, but the financial crash, which hit the United Arab Emirates hard, pushed back the date indefinitely. A truly zero-carbon city proved too ambitious, or maybe too difficult, given the current limitations of renewable energy, so now the aim is for low carbon. Transport within the city will no longer be done solely with the PRTs—instead, electric buses and other mass transit will be included in the mix. Though the first phase of the project—the Masdar Institute of Science and Technology—was completed in the fall of 2010 and opened to students, it's still easy to wonder whether clean-tech companies and expats will be drawn to Masdar, and whether the sustainable city will ever be able to sustain itself. Could Masdar be little more than a desert mirage?

While Masdar may inspire skepticism, it would be a mistake to dismiss the whole project as green folly. That much was clear when I toured the Masdar Institute, the first part of Foster's vision to be completed, on a return visit in 2011. After arriving via the PRT, visitors walk up a spiral staircase to the city's surface. The streets are narrow and sheltered, designed to block the desert sunlight,

FROM LEFT: PIERO OLIOSI—POLARIS; FOSTER & PARTNERS

Foster wants Masdar to look like an ancient Arab city.

while openings in the walls channel a refreshing wind that Masdar officials say makes the city feel as much as 70°F (29°C) cooler than its surroundings. Both features are seen in traditional Arab cities, something that Foster was keen to include in his design. The result is a layout that encourages walking and street life—something rarely seen in modern Middle Eastern cities like Dubai, which have embraced the automobile and vast air-conditioned towers.

That design helps encourage energy conservation—the cooler the city is, the less need for electricity-hogging air conditioning. (Liberal AC use is one of the many reasons Abu Dhabi proper has the biggest per-capita carbon footprint in the world.) The buildings themselves take advantage of green materials, from the sustainable Douglas fir used to build the institute's library to the superstrong plastics that sheathe the laboratories, deflecting sunlight and insulating the interior. Windows have shades angled to avoid direct sunlight, providing light without heat while preserving modesty for the occupants of the residential buildings, in keeping with local customs. There's even a 147-foot-tall (43-meter) wind tower—another high-tech version of something seen in traditional Arabic design—that can funnel even more breezes to the street.

The tower also has glowing LED lights that run down its spine and let Masdar managers know how much energy the city is consuming. Blue means Masdar is within its goal of using 50% less energy than a comparable settlement. Red means it's time to turn off the lights and save energy.

That's the theory—but in practice, those goals aren't always easy for Masdar to meet, at least not by design alone. Martyn Potter, Masdar's director of operations and facilities, noted that most Abu Dhabi citizens are used to keeping their air conditioning as low as 60°F (15.5°C)—encouraged by heavily subsidized electricity—but in Masdar, AC needs to be set closer to 77°F (25°C) to keep within its efficiency targets. With the ability to monitor exactly how much electricity every room in the city is using, Potter can keep the citizens of the Masdar Institute in line. "It's name and shame," he says. "I'm a green policeman."

That might work in a controlled environment—especially one whose residents are working on sustainability. But it demonstrates that even the best green buildings with the best technology work less well when the X factor of actual occupants is included. Some behavior change is necessary—a useful lesson for future green-city planners. The weather can be as hard to predict as the people: The 10-megawatt solar-photovoltaic (PV) field just outside Masdar, which supplies much of the city's power, doesn't work so well when occasional sandstorms muck up the solar cells, reducing their efficiency. The solution was simple—the panels needed to be cleaned regularly with rags—and the experience will be handy for the next Middle Eastern community that tries to scale up solar PV, such as Saudi Arabia, which has its own surprising green dreams.

Will Masdar City ever really develop the authenticity of a real city? It's impossible to predict now, but it's difficult to imagine. The behavioral regulations and controlled design that keep Masdar green might also limit the free and serendipitous qualities that mark a living city—not to mention discouraging potential residents who might not want to follow such a strict rule book. Yet even if Masdar City fails to become everything its planners dreamed when it was launched in 2007, the project will still have enormous value as a living lab for green ideas that can be underwritten with Abu Dhabi's oil money. At a moment when few other countries are taking those sorts of steps, that's important for the planet. "What we're learning at Masdar no one else knows yet," Al Jaber told me. "Masdar will be the global platform to test this technology."

Some of Masdar's technology, like those slightly silly PRT cars—may not have a future. But other ideas, like the wind tower or those sunlight-deflecting windows, will have real value at a time when more than half the world's population lives in cities, a proportion that is growing every day. No one knows the answer to the energy and climate challenges the planet faces, which is why experiments count—even the ones in the desert.

The Big (Green) Apple

New York may not have an environmental reputation, but smart leaders are planning for a greener future for the biggest city in the U.S.

As flat as a pool table and barely a mile wide at its narrowest, the Rockaway Peninsula—a tongue of land that sticks into the Atlantic Ocean at New York City's southeastern corner—is already vulnerable to storm surges and floods. Global warming, with its rising seas and harder rain, will only intensify those threats. That's what has Vincent Sapienza, the city's deputy commissioner for wastewater treatment, so worried. The Rockaway Wastewater Treatment Plant, which processes 25 million gallons (95,000 cubic meters) of sewage a day, sits next to the beach, and its pumps are below sea level. In a major flood, parts of the plant could be submerged, shutting down sewage treatment. "If you lose these pumps, you're done," says Sapienza, standing in the plant's churning basement. "This is a really vulnerable place."

To prepare for climate change—and growth—the city is spending $30 million to raise the pumps and other electrical equipment at the Rockaway plant well above sea level. The overhaul is just one part of New York's groundbreaking PlaNYC—a long-term blueprint to turn the U.S.'s biggest city green in the age of global warming. "This is about making the city more sustainable," says Sapienza.

Though it's caricatured as a concrete jungle, New York is already surprisingly eco-friendly. Thanks to its density and public transit, the city has a per capita carbon footprint 71% smaller than the U.S. as a whole. With more than 8.2 million people calling New York home, the city's infrastructure—its crowded subways, traffic-choked streets, aging water mains—is being pushed past its limits. City planners realize that New York is on track to gain an additional 900,000 people by 2030. If that growth isn't managed properly, the result will be an environmental and economic mess. "New York is growing, and we have to think more effectively," says Rohit Agarwalla, the former director of the city's Office of Long-Term Planning and Sustainability. "We can't just build more power plants. We can't just grow on the edges."

The answer to the question of where the city will put nearly 1 million extra people is PlaNYC. Unveiled by Mayor Michael Bloomberg on Earth Day 2007—and pushed since then with all his considerable political capital—PlaNYC includes more than 120 green initiatives that range from planting 1 million trees to cleaning up every square mile of contaminated land in the city.

Ultimately PlaNYC attempts to chart New York's growth by vastly improving energy efficiency in the city's 950,000 buildings, beefing up public transit, and adapt-

SEAN HEMMERLE—CONTACT PRESS IMAGES

A new aqueduct will connect New York City to pristine upstate water supplies.

ing to the impact of global warming. "If we can solve these challenges here, we can solve them anywhere," says Ashok Gupta, the air- and energy-program director for the Natural Resources Defense Council.

The city started by focusing on what it could control directly. Bloomberg launched a $2.3 billion plan in 2008 to reduce carbon emissions from city-owned properties 30% by 2017 by retrofitting buildings with more efficient lights and better insulation. The payoff is that the city expects to begin saving money through reduced energy bills as early as 2015. On the streets, 33% of the city's 13,000 taxis are hybrids, with more on the way. "The city has made progress on improving what it can control," says Jonathan Rose, a New York architect. "The place where work is really needed is greening all the other buildings in New York."

One area where Bloomberg's green vision has clashed with political realities is mass transit. The subway system is controlled not by the city but by New York State's Metropolitan Transportation Authority. So while PlaNYC includes a call for the subways to be brought up to a state of good repair (a visit to any subway station will indicate they're not there yet), the city doesn't have the power to enforce it. Similarly, the plan pushes projects like the long-awaited Second Avenue subway line on Manhattan's East Side, now under construction. Those multibillion-dollar improvements were to be paid for in part by implementing congestion pricing in Manhattan—charging drivers to enter the most crowded part of the city. As an added benefit, congestion pricing would have helped unclog New York's traffic, which now costs the city $13 billion a year in lost economic productivity and dirties New York's air, which is more polluted than that of any other city in the U.S. besides Los Angeles. "It's an essential idea," says Steven Cohen, executive director of Columbia University's Earth Institute. And one the state wouldn't approve, which cost the city a one-time federal grant worth $354 million. Increases in capital expenditures and operating expenses could result in the transit authority facing deficits. Without a healthy subway system, New York will be hard-pressed to grow, green or otherwise.

The new High Line park in Manhattan is one of many green spaces the city has installed in recent years.

New York's transit struggles are a reminder that even the biggest city in the U.S. can't fully control its environmental destiny. That's true for climate change too; even if New York meets its laudable CO_2 reduction goals, that alone will do little to stop global warming. But the city is ensuring that it will be ready for a warmer world. The Bloomberg administration began by creating a homegrown version of the U.N. Intergovernmental Panel on Climate Change. Those scientists reported that by the end of the century, annual mean temperatures in New York City could increase 7.5°F (4.2°C), with sea levels rising as much as 55 inches (140 centimeters), depending on how fast polar ice melts. "Coastal floods will be very powerful and very damaging," says Cynthia Rosenzweig, a NASA researcher and co-chair of the New York climate panel.

The panel's predictions will fuel the work of New York's Climate Change Adaptation Task Force—a group of city, state, and federal agencies that control vulnerable infrastructure. Though the adaptation plans are in their early stages, the mayor's office is already beginning to prepare the most vulnerable neighborhoods. That puts New York well ahead of any other major metropolis—and certainly the federal government—in taking a dead reckoning of the risks of global warming.

Bloomberg, the green billionaire, won't be mayor forever. (Presumably.) That means PlaNYC, which runs to 2030, will have to remain relevant long after its political patron is gone. But PlaNYC is built to last, even during a recession, because it encompasses far more than just feel-good greenery. Agarwalla, who has studied why Philadelphia declined compared with New York in the 20th century, believes sustainability will be the key to urban success in the 21st century. "We didn't develop this plan out of a desire to be green," he says. "This is crucial for its economic and environmental future."

FROM LEFT: CAMERON DAVIDSON–CORBIS; MEDIOIMAGES—PHOTODISC/GETTY IMAGES

Stockholm has promoted bike riding and its citizens have responded, lessening traffic and pollution.

Why Stockholm Is a Model Green City

Call it a recycling opportunity. After their failed bid to host the 2004 Summer Olympics, Stockholm city leaders decided to turn a would-be sports village in the Hammarby Sjöstad district into one of the world's most successful eco-villages. The practices of powering buses with biogas, recycling rainwater for irrigation, and using organic waste for fertilizer spread to other districts of Sweden's largest city. Today the city's water is so clean that fishermen actually stand on bridges in the central business district, catching fresh salmon and trout.

Stockholm was named the first European Green Capital in 2010. Since then, green innovation has become a pillar of Swedish national competitiveness. With its target to become a fossil-fuel-free city by 2050, Stockholm hopes to turn green into gold by exporting smart power to an energy-conscious world.

Construction has begun at the new Royal Seaport, where a smart grid will allow renewable energy (including solar and wind power) to flow among the homes and offices of residents. Buildings will become "green houses" that not only use but also store green energy and then feed it back into the grid whenever possible. This should enable yearly carbon emissions to be reduced to less than 1.5 tons per person by 2020, vs. the current U.S. average of 20 tons. Ships will be able to plug in and charge up using the onshore electric grid, meaning they can shut off noisy engines, making the harbor area more attractive to live in—and much cleaner.

On transportation, the city is requiring that at least half of all new private cars should be classified as green, while at least 16% of all fuel must be green by 2015. Already a third of Stockholmers walk or bike to work or school, and during rush hours more than three-quarters of the city uses public transport. And 96% of the city's public vehicles qualify as green. The city has also pledged to remove dangerous substances from households—including toxic chemicals like brominated flame retardants and phthalates. On hazardous waste, Stockholm has already cut down levels to just 5.5 pounds per person, or one-third of the Swedish national average.

Large delegations from nearby Copenhagen and Helsinki and places as far-flung as China have become regulars in Stockholm, taking notes on how the city government is building out its grid through public-private partnerships involving Finnish utilities and Swiss engineering titan ABB.

The next step is to export Stockholm's smart energy to the world. Denmark, for example, is connected by underwater cables. There's talk of using such physical connections to enable development of a pan-European energy grid that would theoretically allow all of Scandinavia to export wind and hydropower southward. Swedish historian Gunnar Wetterberg made waves when he called for the five Scandinavian countries to form a United Nordic Federation within the next two decades. There'd be plenty of votes for Stockholm as its capital.

Global Debate

Global warming is a global problem, which is why it has been so difficult to solve. Progress on a great international treaty to reduce emissions has stalled, partly because the U.S. remains divided on climate change. Is there hope for a real solution?

MATTHIAS KULKA—CORBIS

How to Solve the Climate Standoff

The world seemed ready to take action against global warming, but the result was deadlock and higher carbon emissions. Is there a better way?

CLIMATE CHANGE ADVOCATES have had a tough few years. After the triumph that was the U.N. Intergovernmental Panel on Climate Change and Al Gore sharing the Nobel Peace Prize in 2007, expectations were high that the world would finally take action on global warming. But that's not what happened. Despite the election of President Obama, who promised in his campaign to tackle climate change, a cap-and-trade bill to limit carbon emissions stalled and finally died in Congress in 2010. Matters were no better internationally—the 2009 U.N. climate summit in Copenhagen, which was meant to deliver that global deal, was an unmitigated disaster, and things haven't gotten much better since. The crisis-prone global economy has drained public worry away from the environment, even as growing public debt makes it harder for governments to support renewable energy. The climate is getting warmer, but our efforts to slow global warming are ice cold.

That's what made New York City mayor Michael Bloomberg's announcement in the summer of 2011 that he was giving $50 million to the Sierra Club's Beyond Coal campaign a rare bright spot for greens. The Sierra Club, the nation's largest environmental group, successfully stopped more than 150 proposed coal plants from being built over the past decade through the campaign. Bloomberg's money—and perhaps more importantly, the imprimatur of one of the richest and most influential people in the country—is enabling the Sierra Club to bring its war on coal to a new level, preventing untold millions of tons of greenhouse gas emissions from warming the planet.

The real focus of the Beyond Coal program is less about cutting carbon, however, than it is about reducing conventional pollutants that directly affect human health. If we're smart, this approach might be the new way to attack climate change: by identifying actions that can provide a wealth of benefits—including on carbon emissions—rather than simply focusing on global warming alone. That's the message of an influential paper called "Climate Pragmatism" that was published in 2011 by a bipartisan range of thinkers on energy and climate issues. The best way to deal with climate change, as it turns out, is not to deal with climate change, at least not directly.

It sounds a bit confusing—if this is a historic challenge, why not just tackle it head on? The answer is simple: We can't, or at least, we refuse to, as the last few years have shown. There was a clear rise in climate skepticism—especially in the U.S. but also in other parts of

ETHAN MILLER—GETTY IMAGES

Obama pushed climate legislation and clean energy, but met Republican resistance.

Local protests against coal pollution has had more success than the larger global battle against climate change.

the world. A 2010 BBC poll found that 25% of Britons did not believe that global warming was happening, up 10% from the previous year, while a 2010 Gallup poll found that just 53% of Americans saw climate change as a serious threat, down 10 points from the previous year. At the same time, while the science linking carbon emissions to warming is still robust, it remains difficult for researchers to predict exactly how severe climate change will be. And that in turn makes it hard for us to know just how much we should spend to avert that warming. The failure of the global deal is an inevitable consequence of what Roger Pielke Jr., a professor of environmental science at the University of Colorado and one of the authors of the "Climate Pragmatism" paper, calls "the iron law of climate policy." Any climate policy that is viewed as obstructing economic progress will fail—especially in large developing countries that are counting on rapid economic growth to lift citizens out of poverty. Take China, for example: While the country has emerged as a world leader in terms of clean energy investment, its leaders remain reluctant to sign onto any kind of meaningful carbon reductions. The economy comes first, with renewables supplying just a tiny portion of China's overall energy mix. Coal is far more important, with coal imports in China and India slated to grow 78% in 2011.

This means any global carbon cap that would raise the price of fossil fuels significantly simply won't fly in China—or for that matter, in the U.S. But that doesn't mean there's zero willingness to consider the environmental or health perspectives of the energy we use. The developed world has vastly reduced air pollution over the past several decades through ever-tougher regulations on conventional pollutants like soot and acid-rain-causing sulfur dioxide. These are rules that, despite constant industry opposition, remain broadly popular among the public, much more popular than carbon regulations, because the benefit is visible, immediate, and personal.

Developing countries will be no different. Conventional air pollution is a tremendous threat to Chinese growth and public health, as anyone who watched the Beijing Olympics in 2008 knows. Air and water pollution costs China an estimated 4.3% of its GDP each year, and globally, air pollution contributes to an estimated 3 million deaths a year. Any policies or efforts that divert investment from the dirtiest sources—as the Sierra Club is doing with its Beyond Coal campaign—towards cleaner alternatives like natural gas and renewables will benefit public health, while helping the climate as well.

One target could be black carbon, a fancy word for soot, which not only causes serious respiratory problems but also contributes disproportionately to the warming of the atmosphere and especially high-altitude snow cover. (Black carbon can actually settle on white ice, darkening it and causing it to absorb more sunlight and melt faster.) Unlike carbon dioxide, black carbon is relatively easy to control with better engines and cleaner fuels, and tackling the pollutant pays off immediately for health and the climate as well. It's even bipartisan: In 2009 the

FROM LEFT: ANNA WEBER—WIREIMAGE; BILL CLARK—ROLL CALL/GETTY IMAGES

staunchly Democratic senators John Kerry and Barbara Boxer joined with the Republican climate-change skeptic James Inhofe to co-sponsor an effort to investigate ways to reduce black carbon.

At the same time—especially for developing countries—those alternatives likely need to be economically viable. For most of the world, the opposite is true, which is why more than 1.4 billion people lack virtually any access to electricity. That's an astounding figure, but one that rarely gets the attention it deserves. Lack of electricity impacts public health—try running a modern hospital without any power—and retards economic growth. If we want developing nations to be better prepared to deal with the effects of climate change, or just about any other threat, we need to get them wired.

The challenge will be to develop low-carbon alternatives that can compete with fossil fuels on price. (Subsidies are limited—already, even ultra-green countries like Germany are cutting back aid for renewable power because of the rising price tag.) If alternatives are going to win they need to get a lot cheaper and a lot more efficient, and that's going to require vast increases in the amount of basic R&D spent on energy. The American Energy Innovation Council, a heavyweight lobbying group that includes Bill Gates, has suggested that the U.S. should increase funding for energy research around $3 billion a year to at least $15 billion annually.

Lastly there's the pressing need to adapt to climate change. It seems like a no-brainer, but we need to think a little harder about what adaptation actually means. There's an assumption that we can actually separate adapting to climate change from the act of preparing for any natural disaster or extreme weather. In reality, though, parsing the two is nearly impossible—we still can't assign blame for specific weather events—and absolutely pointless. The climate adaptation assistance that rich nations are sending to the developing world is almost totally drawn from the existing budget for foreign aid.

A hurricane will create havoc for an unprepared population whether the storm has been strengthened by carbon emissions or not. Countries need to be prepared for all the stresses the future will bring—from extreme weather to higher energy prices to infectious disease. The watchword should be resilience—creating societies that can bounce back from anything—and the best way to do that is through continued economic development.

Nationally and internationally, climate politics are deadlocked, even as carbon emissions keep rising and the weather keeps getting weirder. What could work is an oblique approach to climate change, one that sidesteps the roadblocks by taking advantage of popular, no-regrets actions that are worth doing even if global warming wasn't real. It's not as simple or as elegant as one global deal—but it might actually be what we need to survive.

Congress and Climate Change

It's hard to remember, but both candidates in the 2008 Presidential election believed in climate change and promised to deal with it in office. Barack Obama and John McCain were even going to use the same general method: a carbon cap-and-trade program that would increasingly restrict U.S. greenhouse gas emissions. After Obama was elected—with a Democratic majority in both chambers of Congress—the assumption was that climate legislation was a done deal.

That's not how it turned out. Though the House of Representatives managed to just barely pass a cap-and-trade bill in 2009, the Senate was never able to even vote on legislation. When Democrats were routed in the 2010 midterm elections, any hope that climate legislation would be passed died as well—even as global efforts ground to a halt too.

What happened? The events of "Climategate" in 2009—which saw thousands of hacked messages from climate scientists published on the Internet—and mistakes in the U.N. global warming reports damaged public belief in climate science. The Republican party—led by longtime skeptics like Senator James Inhofe of Oklahoma—turned lockstep against even the idea of fighting climate change. But most of all, the lengthy economic crisis sapped the public's will to take steps now to avoid a warmer future. Climate change is scary, but for most Americans, imminent unemployment is a lot more frightening.

Inhofe helped stop climate legislation from passing.

TIMOTHY HEARSUM—GETTY IMAGES

Solar Eclipsed

These should be boom times for U.S. makers of solar gear, but China is running away with the business. Is this foul play?

IT SEEMS LIKE BOOM TIMES FOR U.S. SOLAR. With demand skyrocketing, about $11 billion worth of solar-power gear is set to be installed in 2012, and more than five times that amount is coming down the pike. Solar is employing 100,000 Americans, a number that rose by 7% in 2011 even as overall employment barely grew at all.

But even as solar power thrives in the U.S., many think the business could be growing faster—and creating more American jobs—if it weren't for alleged foul play by China, the country's biggest solar rival. "The Chinese are eating our lunch," said Michigan Representative John Dingell, a Democrat, during a congressional hearing on renewable-energy funding. The feeling resonates with politicians who fear the U.S. is

The solar industry is growing rapidly around the world—but which country will dominate?

losing its edge because of unfair trade practices abroad, especially in China. In his State of the Union speech in January 2012, President Obama vowed to take action "when our competitors don't play by the rules."

But despite the ripe political climate for erecting trade barriers, the solar industry is split about the merits of protection. The division is simple. If you're a customer buying solar panels or running a business that installs or services them, you're doing well. But if you make solar modules—especially in the U.S.—your balance sheet looks ugly. That's because solar power is getting much cheaper: Prices for modules have dropped 40% in recent years, and costs seem likely to continue falling. "The good news for solar is that it's rapidly getting less expensive," says Kevin Lapidus, a senior vice president at the solar-services company SunEdison. "Eventually we'll sell solar the same way we sell anything."

But some U.S. manufacturers believe cheap imports from Chinese panel makers, which receive billions of dollars in aid from Beijing, are causing the nosedive in solar prices. As a result, China now produces three-fifths of the world's solar panels—a proportion that is likely to increase. "Western manufacturers cannot survive this," says Ben Santarris, spokesman for the U.S. arm of SolarWorld, a major German panel maker.

In 2011, those concerns prompted SolarWorld, on behalf of seven solar manufacturers, to file a complaint of unfair trade practices by China. The Obama Administration has punted on the issue in the past, partly out of fear of igniting a trade war with Beijing, which has already threatened retaliatory action. But there are other qualms. While tariffs might help some U.S. manufacturers in the short term, both consumers and the larger domestic solar industry would likely suffer if the resulting higher prices hampered demand. A recent study commissioned by the Coalition for Affordable Solar Energy (CASE)—a trade group of solar companies that oppose tariffs—found that a 100% tariff on imported modules would result in a net loss of as many as 50,000 jobs in the U.S. over the next three years and would cost consumers as much as $2.6 billion; a 50% tariff could eliminate up to 43,000 jobs and cost consumers as much as $2.3 billion. "The analysis makes it clear that tariffs on polysilicon solar cells would be devastating for American workers," says Jigar Shah, president of CASE.

That might sound surprising; after all, tariffs are supposed to protect domestic workers. And a coalition of U.S. solar manufacturers that support tariffs noted that the study was "highly speculative" and depended on optimistic projections for solar growth in the years ahead. But installers and service providers point out that manufacturing jobs make up less than a quarter of the roughly 100,000 jobs in the U.S. solar industry, with far more found in maintenance, installation, sales, and service. In that way the solar industry is like the U.S. economy as a whole. Despite all the lip service politicians give to factories, less than 10% of American jobs are in manufacturing, down from around 30% in 1950. "The jobs in this industry are increasingly found outside of manufacturing," says Lapidus.

A Chinese worker at world-leading Suntech assembles new solar panels.

Nonetheless, in May 2012 the Commerce Department decided to impose antidumping tariffs of more than 31% on Chinese solar panels. It's one of the biggest antidumping decisions in U.S. history, covering one of the fastest-growing categories of imports from China. The ruling, which could be challenged at the World Trade Organization, will likely have a major impact on Chinese solar manufacturing. (The U.S. bought $3.1 billion worth of Chinese solar cells in 2011, which gives China more than half of the U.S. market for panels.) Chinese officials responded angrily to the decision, noting that it was hypocritical for the U.S. to urge China to develop renewable energy and reduce its carbon emissions while slapping tariffs on imported solar panels. "U.S. tariffs will hurt both countries because China imports a large amount of raw materials and equipment from the U.S. to produce solar panels, and it exports such goods to the U.S.," said Shen Danyang, a spokesman for China's Commerce Department.

The new taxes have had an impact on Chinese solar companies, which were already struggling with a highly competitive—and oversupplied—local market. Some firms may end up shifting some of their production to the U.S. or other countries in an effort to avoid the tariffs, just as Japanese car makers like Toyota opened up American plants under pressure from trade disputes in the 1980s. But the pay gap between American and Chinese workers in the solar manufacturing industry today is far larger than the gap between American and Japanese auto workers some 30 years ago, which may limit the ability of Chinese companies to relocate profitably. It's more likely that the U.S. tariffs will lead to a response from the Chinese government—there's already a battle over wind turbines—and the solar conflict could lead to something closer to a full-scale trade war between the two countries. That would have negative ramifications for the solar industry—and the global economy as well.

Still, environmentalists have sold the American people on the idea of green jobs—and if all those jobs end up going to other countries, support for climate action might evaporate as well. The question is whether China's protections would give it a leg up in other areas of solar in which the U.S. remains competitive. A recent government study suggests American solar companies still have an innovation edge over their Chinese counterparts. Of course, environmentalists say the focus of any solar policy should be the planet, and that means making solar power cheaper faster. "China's focus on renewable energy and high technology is here to stay," wrote Melanie Hart, an energy analyst for the Center for American Progress, in a recent research note. "That can be a great thing for the U.S." as it seeks to create a greener economy. The real war over solar isn't the U.S. vs. China—it's solar vs. fossil fuels. And that victory is still up for grabs.

FROM LEFT: QILAI—IN PICTURES/CORBIS; PARTH SANYAL—REUTERS/CORBIS

How India is Reinventing Solar

In 2009, when policymakers in New Delhi set a goal to produce 20,000 megawatts of solar energy by 2020, few gave India more than a slim chance. All the world's solar-savvy countries put together were generating that much solar power at the time, and India was contributing virtually nothing. But today, with acres of land in its arid, sun-drenched northwest area carpeted with thousands of gleaming solar panels, analysts say India will exceed its target. In just one year, funding for solar projects in India increased seven-fold, from $600 million in 2010 to $4.2 billion in 2011.

How did India catch up? First the global price of solar panels and modules that turn sunlight into electricity plummeted 30% to 40%, triggered by a massive expansion in China—home to the world's leading panel makers—and tepid demand from Europe. While this brought doom to American manufacturers unable to compete with China's prices, it proved transformative for the industry by making solar infrastructure more accessible. Germany added a record 7.5 gigawatts of panels in 2011, more than double the government's target. In the U.S., grid-connected solar installations in the third quarter of 2011 grew 140% over the previous year. Indian developers too decided to join the party.

Driven by its ambitious new solar policy, the Indian government agreed to buy solar power at 17.91 rupees (36 cents) for a kilowatt hour. (India's coal-generated energy costs 3 to 4 rupees.) To the government's surprise, it received an overwhelming response from developers. That's when India set up a reverse auction process, making developers compete for its business. "The Indian experiment has been very successful," says Tobias Engelmeier of Bridge to India, a New Delhi–based consultancy.

While solar energy is getting more attractive, what's tilting the Indian energy market further is that coal is becoming more expensive. By 2017 domestic coal production in India will meet only 73% of demand, making imports imperative. This dramatic fall in solar prices has, however, raised some questions: Are these projects, almost too good to be true, financially feasible? Analysts warn a weeding-out process is in the offing. Nevertheless, solar "looks like it will be a significant source of energy" going forward, says Alan Rosling, co-founder of Kiran Energy, a solar developer whose story mirrors India's own growth trajectory. The company now owns plants sprawled across 125 acres and has bagged contracts for 75 megawatts of solar power. Setting the bar high, Rosling says solar will have "truly arrived" in India when developers can sell it to anyone at a competitive price without relying on the government. *—By Niharika Mandhana*

A laborer connects solar panels in the Indian city of Kolkata. India has a need for electricity—and solar could supply it.

Trouble Spots

From depleted forests to dying reefs, distress signals dot the globe. Even in the U.S., with its relatively clean environment, excessive carbon emissions fuel global warming.

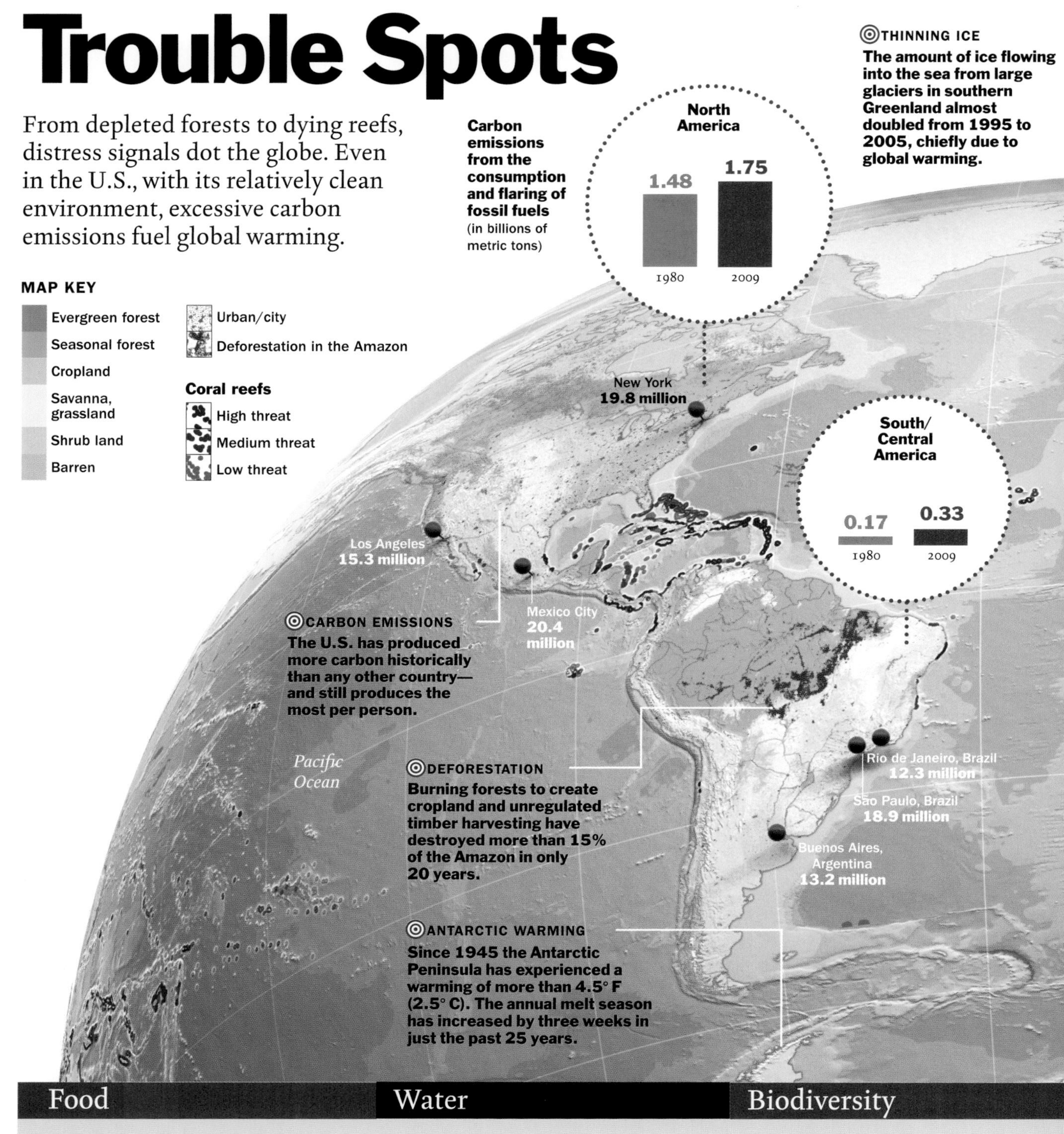

Food

The green revolution helped feed developing nations in the latter half of the 20th century. But hunger continues to plague poorer countries, especially in Africa, as badly managed agriculture often leads to soil salinization and degradation.

Water

As more of the limited amount of fresh water is used each year, unequal access to supplies could produce competition and conflicts among nations. If polar ice caps continue to melt down, a major problem of the 21st century may be the rising tides of seawater.

Biodiversity

Destruction of forests and rainforests has helped cause the worst spasm of extinctions since the dinosaurs fell victim to an asteroid impact 65 million years ago. A 2006 report linked the extinction of frog species in Central America to the emission of fossil fuels.

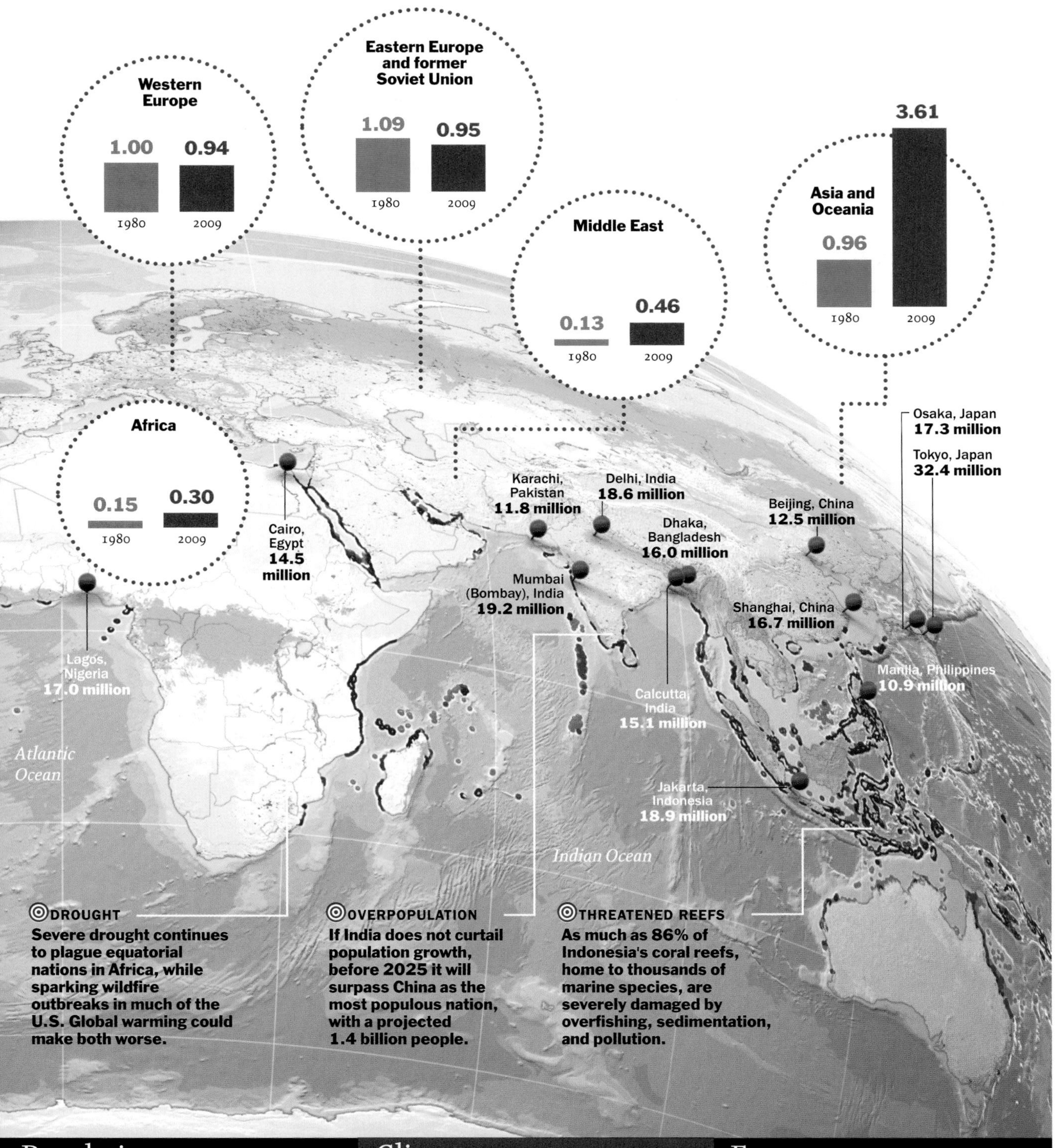

Population

Life expectancy is increasing around the globe except in Africa, where AIDS and other infectious diseases have taken a toll. Lower birth rates will start to level off population growth by mid-century.

Climate

The phaseout of chemicals called chlorofluorocarbons, achieved by a 1989 global pact, will help reduce the hole in the ozone layer. But the burning of fossil fuels will lead to hotter times in the future.

Energy

Humankind's continued reliance on fossil fuels that emit carbon dioxide is extremely harmful to the planet's climate. The search for alternate fuels will be a dominant theme of 21st-century science.

SOURCES FOR MAP Land use: NASA/Boston University Department of Geography; urbanization: NASA Visible Earth City Lights; U.N. Population Fund, 2000; Amazon deforestation: ActGlobal.org/Instituto Socioambiental; coral reefs: World Resources Institute: Reefs at Risk; carbon-dioxide emissions: Energy Information Administration; trouble spots: AP; U.N. Environment Program; Global Warming Early Warning Signs 1999; World Resources Institute

Green Heroes

Some are scientists and some are activists. Some run businesses and others run parks. But the men and women highlighted in these pages all have one thing in common: They care about the planet, and they're ready and willing to fight for it.

JASPER JAMES—GETTY IMAGES

Protectors Of the Planet

When it comes to cleaning up the planet, a few smart people with a few good ideas can often make all the difference. Here are some of the best.

THOMAS MEYER—OSTKREUZ

Citizens

THE RESIDENTS OF VAUBAN

We know cars are terrible polluters and emit a huge chunk of the world's greenhouse gases. But how many people are prepared to give up their car for the good of the planet? In Vauban, a district on the outskirts of Freiburg, a city in southwestern Germany, the answer turns out to be quite a few. For the past decade, cars have been banned in most of Vauban: no home garages, no street parking, and a charge of some $30,000 for a space in one of two multistory car-parks. The impact has been dramatic: The car-ownership rate among the 5,000 residents has plummeted.

Perhaps the most surprising thing about the district's experiment in car-lessness is that it was local residents who pushed the idea. Almut Schuster belongs to a car-sharing club so on the few occasions that she needs a car, she teams up with other residents and shares a lift. She also lives in an apartment in which the water is heated by a rooftop solar panel and the power comes from a supplier who uses renewable sources such as a local wood-chip–fired power plant. "There are many options for using renewable energy at home," Schuster says. "We all share this planet and we need to be conscious of how we live and what we eat."

Activist
DAVID SUZUKI

Born in British Columbia's Japanese-Canadian community, Suzuki was interned with his family as a young boy during World War II. Unjust, yes, but the camp was in beautiful territory. Nature's glory and mystery imprinted early on him; he grew up to become Canada's premier young geneticist, an award-winning bench scientist who became a professor at the age of 33.

But something made him restless with his fruit flies, and before long he'd embarked on a second career, creating nature documentaries for television and radio. The best-loved of them, a TV series called *The Nature of Things*, began its run in 1979 and has aired in some 50 countries around the world, making Suzuki a kind of terrestrial Jacques Cousteau.

Ecological science holds that everything is connected. If so, Suzuki has become one of the crucial hubs in the cultural ecology of our strained earth. Biologists talk about keystone species essential for the proper function of an ecosystem; Suzuki is a keystone guy.

Royal
PRINCE MOSTAPHA ZAHER

Young Prince Mostapha Zaher's first kill was a swift, one of those fast-flying birds that swooped through the palace grounds every afternoon in the long Kabul summer. He shot it with his air rifle, and when he brought the bloodied trophy home to his father he was hailed as a great hunter—one in a long line of hunters. But his grandfather, Afghanistan's King Mohammed Zahir Shah, was less impressed. "He scolded me," chuckles Zaher. "Swifts are precious creatures of the air, and if you can't eat them, you don't hunt them."

The message stuck, and while Zaher continued to hunt, it was always with his grandfather's lessons in conservation in mind. Another legacy remained as well: the late King's desire to turn the royal hunting grounds of the Ajar Valley, a spectacular mountain refuge in central Afghanistan, into a preserve open to all Afghans.

That dream was revived when the King and his family—deposed and exiled to Italy in 1973—returned following the fall of the Taliban. Zaher took up the job of director of Afghanistan's newly formed National Environmental Protection Agency in 2004.

Since then he has worked to rewrite the nation's environmental laws, enshrining in the constitution an act that declares it the responsibility of every Afghan citizen to "protect the environment, conserve the environment, and to hand it over to the next generation in the most pristine condition possible."

CLOCKWISE FROM TOP RIGHT: MARTIN HARTLEY; BRIAN SMALE; MARTIN HARTLEY—REPORTAGE/GETTY IMAGES; CARMEN SCHMID

Explorers

PEN HADOW, MARTIN HARTLEY AND ANN DANIELS

Scientists know that global warming is thinning the Arctic ice cap. They're just not sure how fast. So in 2009, three British explorers set out from northern Canada to gather data that will help scientists assess how global warming is affecting sea ice. During their 73-day, 270-mile trek, Ann Daniels, Pen Hadow, and Martin Hartley took 1,500 measurements of sea-ice thickness and recorded 16,000 observations on everything from snow distribution to the size of cracks in the ice. "[Explorers have] mapped the world's surface," says survey leader Hadow. "The next phase is getting to places that are too hazardous for standard scientific operations and extracting raw information."

Even for an Arctic veteran like Hadow—in 2003 he became the first person to trek from Canada to the North Pole without resupply—getting that information would prove unexpectedly arduous. They would spend up to five hours a night drilling in temperatures as low as -94°F (-70°C). Crossing the hostile Arctic terrain took a heavy physical toll on the team. Hartley shed 35 pounds (16 kilograms), and almost lost a big toe to frostbite. Hadow and Daniel's feet and hands suffered temporary nerve damage from the cold. But their sacrifices are paying off for scientists around the world—and for the rest of us as well.

Designer

VALERIE CASEY

In the spring of 2007, Valerie Casey started to sketch out a Kyoto Protocol of design: a set of measures and targets that would put sustainability at the heart of the industry. The big players rushed to sign Casey's call to action, now known as the Designers Accord. Design powerhouses like IDEO and software titans like Adobe are among the thousands of design firms, corporations, colleges, and individuals worldwide that have adopted the document. Signatories agree to follow five green guidelines, including reducing their carbon footprint each year, educating staff in sustainability, and discussing environmental impact issues with every client. At town-hall meetings held regularly across the U.S. and in online forums, designers and firms trade advice on topics from choice of materials to carbon-auditing to cutting unnecessary packaging. It's design for a better planet.

Physicist
OLGA SPERANSKAYA

When the hammer and sickle finally fell in the Soviet Union two decades ago, hundreds of thousands of tons of obsolete pesticides and other chemicals remained. Stored in torn bags and collapsing sheds, the chemical cocktail was allowed to seep into groundwater and from there it passed into the surrounding animal and human populations. The problem had grown so bad, says Russian activist Olga Speranskaya, that a new type of hammer had to be forged, something with which to bang away at the government, "to push the authorities to clean up these sites." The physicist has been pounding Moscow since 1997, demanding it secure stockpiles of chemicals such as DDT—long banned in the West—and help clean up the enormous mess left by the Soviets.

But Speranskaya hasn't just been on the attack. Through her work with Moscow's Eco-Accord Center for the Environment and Sustainable Development, an independent environmental watchdog, she has also educated thousands of people about the dangers chemicals pose, and has brought dozens of activist groups together to make their voices louder. "The environment is beyond any political issues," she says. "We need to continue working—to fight this legacy and to not allow the authorities to make it even bigger."

Scientist

DAVID KEITH

"It's about tools." That's how David Keith sums up his contribution to climate-change research. It sounds quite modest, and Keith, a balding but boyish professor at Harvard University, can be disarmingly modest. But the solution he is researching is immodest in the extreme: geoengineering. The fact that geoengineering schemes—intervening in the climate by shielding the earth's surface from trillions of watts of sunshine, or sucking billions of tons of carbon dioxide out of the atmosphere—have started to get serious attention from policymakers is in large part due to Keith's work. "While he's got informed and strong opinions," says Bill Gates, who relies on Keith for advice on climate issues, "he's also incredibly open-minded, pointing out the unknowns in his opinions and just as readily pointing out the merits of others' opinions."

That balance helps. Keith is keen to stress that geoengineering is no alternative to reducing carbon emissions, but insists that it should be researched as a possible aid. While it excites him as a technologist and fascinates him as a policy wonk, he remains an analyst, not an advocate. The thing about tools, he says, "is not that you have to use them: it's that you have to understand them."

Filmmaker

DAVID ATTENBOROUGH

There are plenty of other missionaries for the environment, of course. But what distinguishes David Attenborough is that boundless, schoolboyish enthusiasm, the infectious joy of discovering the infinite variety of life. It all began over 50 years ago, with Zoo Quest, and reached its apogee in the 13-part BBC series *Life on Earth*, reckoned to have been watched by 500 million people. He is probably the best-known broadcaster in the world. He has been knighted and fêted, of course (and honored by having a wondrously weird New Guinea spiky anteater named Attenborough's long-beaked echidna). But his true legacy is the sense of wonder that he has brought to people all over the globe at the astonishing ingenuity of the life forms with which we share this increasingly crowded space.

Royal

PRINCE CHARLES

The royal radical has been promoting environmental ideas for most of his adult life. Some of his notions, which once sounded a bit daft, were simply ahead of their time. Take his views on farming. Prince Charles' Duchy Home Farm went organic back in 1986, when most shoppers cared only about the low price tag on suspiciously blemish-free vegetables and unnaturally large chickens piled high in supermarkets. The Prince's farm supplies produce to Duchy Originals, a firm he set up in 1992. In what he calls a "virtuous circle," the company markets organic products such as cookies and soups made from the produce grown by his own farm and from ingredients sourced from other suppliers using farming methods that protect the countryside. His warnings on climate change proved prescient, too. Charles began urging action on global warming in 1990 and says he's been worried about the impact of man on the environment since he was a teenager. Charles may seem like a throwback, but he's on the cutting edge of conservation.

CLOCKWISE FROM LEFT: ALEXANDER GRONSKY—AGENCY.PHOTOGRAPHER.RU; HAN LEE DE BOER—CAMERA PRESS; REX/REX USA; EWAN NICHOLSON

Inventor

BINDESHWAR PATHAK

Persistent discrimination against low-caste toilet scavengers is only part of India's serious sanitation issues. Today, some 110 million Indian households remain without access to a toilet and 75% of the country's surface water is contaminated by waste. More than half a million children die each year from preventable sanitation-related diseases. Bindeshwar Pathak realized the only way to solve the problem was to develop a clean method of human-waste disposal that would be cost-effective for the average Indian household and would rid the country of the practice of scavenging. He developed the technology for a new toilet and founded the nonprofit Sulabh Sanitation Movement to bring his creation to those who needed it the most.

Pathak's $15 twin-pit toilet can be installed in any village, house, or mud hut. While one pit is in use, the other is left covered. Within two years, the waste in the covered pit will dry up, ridding itself of pathogens, so that it's suitable for use as fertilizer. The toilets use 0.4 gallon (1.5 liters) of water per flush, as opposed to the 2.6 gallons (10 liters) required by conventional toilets. They also eliminate the need for manual scavenging, so Pathak's NGO—now called the Sulabh International Social Service Organization—also runs rehabilitation programs for out-of-work scavengers, teaching them the skills they need to find new jobs.

Activist

ANNIE LEONARD

Not many people can imagine spending an entire evening listening to stories about garbage and be completely mesmerized. That's because they haven't met Annie Leonard. She has been relentlessly explaining the absurdity of our throwaway culture for decades. Leonard knew her story needed to reach as many people as possible to make a real difference. So, in 2007, she made it viral through an infectious online film called *The Story of Stuff*. Within six months, more than 3 million viewers from around the world watched the film. *The Story of Stuff* effectively and often humorously explains where all our stuff comes from, what resources are used to create it, whose lives are affected during its production, and where it goes when we discard it. While this all sounds familiar enough, it's Leonard's poignant questions and provocative truth-telling that help us see the profound stupidity of this system. Which is why when Leonard talks trash, people cannot help but listen.

CLOCKWISE FROM TOP LEFT: BHARAT SIKKA; PAUL SAKUMA—AP/CORBIS; ELISA HABERER; PHILIP HOLLIS

Pollution Monitor

ZHAO ZHONG

Gansu province in northwestern China is a beautiful and fragile place. Water supplies are prone to overuse by industry and agriculture. Rings of hills trap noxious emissions over cities like provincial capital Lanzhou.

But one man is helping Gansu cultivate a respect for its natural gifts. Zhao Zhong set up Green Camel Bell (GCB)—the province's first environmental NGO—with a few volunteers in 2004. Named for the bells used on traditional camel trains, GCB raises awareness about green issues, monitors polluters, and advocates new policies.

Multinationals have taken note. "Companies that are causing pollution feel pressure from the water-pollution map," says Zhao. "They are forced to take immediate and more open measures to solve the problem." Everyone can drink to that.

Chef

ALICE WATERS

It has been slow progress for all environmentalists, but Alice Waters has more right than most to be frustrated. She just wanted people to eat stuff that tastes better. And it wasn't like she was simply making claims that local, organic food tastes great. She was proving it every day at Chez Panisse, the Berkeley, Calif., restaurant she opened in 1971—a restaurant so good that it doesn't even have a menu. You eat what Waters found at the markets that day, and you like it. You really like it.

While Waters' restaurant and cookbooks are credited with launching the locavore movement in the U.S., her Edible Schoolyard project has gone one step further. It encourages students in Berkeley to help grow and shop for their lunches. It has shown results and has spread to other cities. "Remember when Kennedy put physical fitness in schools?" Waters asks. "We had to exercise four times a week, and we all went for it. We need that kind of passion. Going into public schools and teaching [children] about the consequences of the food that they eat can have remarkable results."

Photographer
YANN ARTHUS-BERTRAND

Soaring high above the earth, Yann Arthus-Bertrand takes aerial photographs that offer an intoxicating perspective on our world. But the photographs also act as something of a visual ecology lesson: Our planet is fragile and threatened by ominous forces. To overcome pollution, deforestation and climate change will require concerted action from we humans, the ones looking down on all this.

Arthus-Bertrand began shooting from the sky three decades ago. His first photographs were taken from hot-air balloons. His UNESCO-backed "Earth from Above" project has been seen by more than 120 million people as a touring exhibition; as a lavish coffee-table book it has sold more than 3 million copies in 24 different languages. "I try to show our impact on our planet," he says. "From the air, you can see the earth's wounds."

Many are new traumas, like the outflows of waste from tar-sand extraction in Canada, toxic landfills in Dakar, Senegal, or the passage of an icebreaker through dappled, melting Arctic floes. "We don't realize the incredible imprint of man," he says. "Sure, life is good now. But we are exhausting our resources. There's not enough fish, not enough wood, not enough land. We have to do better with less."

Philosopher
SHERI LIAO

For Sheri Liao, the solution to the problems caused by China's breakneck modernization can be found in centuries-old wisdom. Before launching Global Village of Beijing (GVB)—one of China's earliest environmental-advocacy groups—in 1996, Liao taught philosophy. As a researcher at the Chinese Academy of Social Sciences, she first came across the idea of "adaptation to nature" in a paper. "I was absolutely shocked," she recalls. "For decades we were told [by Chairman Mao] that our role is to conquer nature. Who'd have thought we could live with it in peace?"

Liao was helped by the fact that the birth of GVB coincided with China's economic takeoff in the mid-'90s. The group became active in Beijing neighborhoods, raising environmental awareness on the local level. But it has expanded, and it is now involved in everything from promoting recycling to encouraging building managers to reduce electricity consumption. The idea, as Liao puts it, is to promote "a life of harmony"—an approach that preaches balance between the body and the mind, the individual and society, and people and the planet.

CLOCKWISE FROM BOTTOM LEFT: JEROME BONNET—CORBIS OUTLINE; ELISA HABERER; CANDACE FEIT; PETER ESSICK-AURORA PHOTOS; AHIKAM SERI—PANOS PICTURES

Campaigner
MARC ONA

Marc Ona's campaign to halt a gigantic mining project in the heart of Gabon's rain forest has earned him a few weeks in jail, regular harassment by police, and, when his landlord grew nervous about provoking government ire, eviction from his home. But Ona's determination has also generated enough local and international attention to shame Gabonese officials into vastly scaling back the project—and possibly to derail the mine altogether. In Ona's crosshairs is a $3.5 billion iron-ore project that initially covered 3,000 square miles (7,700 square kilometers) of the Ivindo National Park, part of the world's second largest rain forest. Through Brainforest, an ecological organization he founded in 1998, and a network of Gabonese green groups, Ona denounced the planned mine, hydroelectric dam, railroads, and deepwater port.

Ona plans to use the $150,000 he received as a Goldman Environmental Prize winner in 2009 as seed capital for microbusinesses begun by local entrepreneurs, many with a sustainability bent. If he can demonstrate the region's economic richness without plundering it, he says, large-scale projects that damage the environment will at least have some competition in the future.

Paleoclimatologist
LONNIE THOMPSON

Until Thompson came along, almost all ice cores were taken from relatively accessible polar regions. He decided to drill where others were not venturing: in the glaciers that crown tropical mountain ranges in places like Ecuador, Nepal, and Tibet. It was intensely challenging work, but the mountain ice cores he has analyzed have added immeasurably to our understanding of the earth's climate by broadening research to the tropics, where 70% of the world's population resides. But Thompson's most lasting accomplishment may be his powerful eyewitness record of climate change: The rapid retreat of alpine glaciers over the past few decades is one of the most unequivocal signs of global warming in action.

Entrepreneur
SHAI AGASSI

Shai Agassi is part scientist, part visionary, with a lot of salesman thrown in. And he thinks big about the future of electric cars.

Agassi sees Israel as the perfect testing ground for a network of electric cars to be built and serviced by his company Better Place. It's the right size—150 filling stations will cover the country—it has closed borders, and there is an added incentive: Some of the world's top oil producers are unfriendly to Israel. One of Agassi's innovations is to charge users not for the car or battery, but for the electricity they consume, much as cell-phone companies profit from how much customers talk. At current prices, electric cars will be far cheaper to run than gas engines, Agassi says, and will produce zero carbon. "We environmentalists made a mistake," he says. "We ask people to pay more to be green, and we should ask them to pay less." If he can pull that off, Israel could be the first nation to junk its gas-guzzling cars altogether.

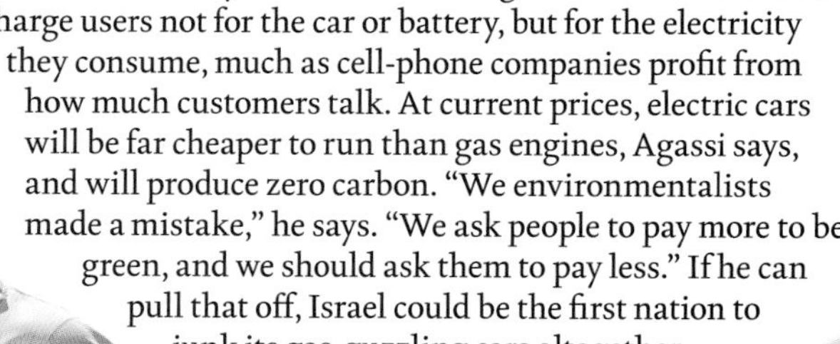

Extreme Measures

What if we can't reduce carbon emissions—and what if climate change turns out to be worse than anyone expects? The only answer might be geoengineering: directly altering the climate to cool the planet. But the cure could be worse than the disease.

GEORGE DIEBOLD—BLEND IMAGES/CORBIS

How to Control the Climate

Geoengineering offers a cheaper way to deal with global warming, but these radical techniques are likely to have unpredictable side effects.

IF THE WORLD IS GOING TO COME TO GRIPS WITH the climate change crisis, it needs to drastically reduce greenhouse gas emissions in the decade ahead, a mission that will require remaking the way we produce and use energy at the likely cost of trillions. Or, maybe, we can just change the color of the sky. Nathan Myhrvold at Intellectual Ventures, a Seattle-based venture capital firm and think tank founded by Microsoft's Paul Allen, has devised a plan to pump around 100,000 metric tons of sulfur aerosol particles into the sky every year, launched into the air using 18-mile-high, 2-inch wide tubes held in place by balloons. Once in the stratosphere, the sulfur particles would scatter incoming sunlight, bouncing some of it back into space and reducing the amount that reaches the earth's surface. Computer models indicate that Mhyrvold's method, carried out indefinitely, could likely keep the planet cool even if carbon emissions keep growing—though there would be some rather noticeable side effects, like a more yellow, hazier sky. And it would cost just $20 million to set up and $10 million a year to run—a miniscule fraction of the likely price of transforming the global energy system.

Mhyrvold's proposal—which he has called the StratoShield—is a type of geoengineering: an attempt to directly control the climate. Long a concept that was kept in the shadows, considered too radical even to try, geoengineering is gradually becoming more and more prominent in climate change discussions. That's in part a reflection of the world's near-total failure to curb carbon emissions, even as global temperatures continue to rise and climate models warn of an ever more fright-

IMAGE SOURCE—GETTY IMAGES

ening future. Geoengineering advocates present their concepts as the planet's ultimate Plan B, a backup that could be deployed quickly if climate change really begins to get out of hand.

But while the economics of geoengineering are tempting, even those who work in the field worry that the existence of an emergency button might sap the will of governments to do the hard but necessary work of reducing emissions. And while climate models are becoming more accurate, we may still know far too little about the planet to begin messing with the global thermostat, which could produce side effects as dangerous as the problem itself. "We are overdue to research this," says David Victor, the director of Stanford University's Energy and Sustainable Development Program and an advocate of geoengineering work. "But there needs to be a very high bar for how much care we take."

Geoengineering breaks into two broad categories: carbon dioxide removal (CDR), taking carbon directly out of the atmosphere and storing it, and solar radiation management (SRM), artificially blocking sunlight to cool the planet. The first, CDR, is already a part of our climate management schemes—forests, after all, suck in and store billions of tons of carbon, as does the deep ocean. This side of geoengineering research is relatively uncontroversial—few people would object to adding more trees or preserving the ones we have in an effort to reduce the amount of atmospheric carbon warming the planet. Most of the new geoengineering techniques that are being developed—like so-called artificial trees that chemically capture CO2 directly from the air and store it—shouldn't produce unexpected side effects. (One

CDR method—seeding the ocean with iron to encourage plankton blooms, which will then absorb carbon—has worried activists who fear the impact on marine systems.) Protecting forests—especially tropical rain forests—also helps remove carbon from the air. Essentially we'd be trying to counteract the harmful geoengineering we're already doing when we pump tens of billions of tons of carbon into the atmosphere through the burning of fossil fuels. "Conceptually it's all the same as planting a forest," says Ken Caldeira, an atmospheric researcher at the Carnegie Institution of Science at Stanford.

Cost, however, is the main limitation on CDR techniques. There are already agreements in place to try to limit deforestation for climate reasons, but they've only been partially successful because the market for wood products and the drive to clear land for agriculture is still strong. Caldeira believes that the only way CDR would be feasible is through a carbon price—a high one,

Myhrvold has big dreams for a geoengineering scheme.

close to $100 a ton in the form of a tax or emissions trading—that would make it financially viable to operate expensive artificial trees. But that would negate the biggest attraction of geoengineering for policymakers: its bargain-basement cost. "A carbon price doesn't look like it's coming on the horizon," he says. And even if we get one, if we're willing to spend that much money, we can just fight climate change the old-fashioned way.

That's where SRM, or solar radiation management, comes in. SRM embraces a broad range of possible actions, from the ambitious—constructing a gigantic suborbital mirror to bounce back sunlight—to the innocuous, like painting highways and roofs white to reflect sunlight rather than absorb it. But the most viable SRM techniques—the ones that excite would-be climate Dr. Strangeloves—involve injecting aerosol particles into the stratosphere, where they could reduce the amount of sunlight actually heating the Earth. Global temperature is a product of how much sunlight reaches the planet and how much is retained through the greenhouse effect. As we've added carbon to the atmosphere, we've strengthened the greenhouse effect, so more of that solar energy is retained in the atmosphere. Block some of that sunlight and you could keep the planet cool even as the greenhouse effect intensified due to increasing carbon emissions.

Scientists know that it will work—when the volcano Mt. Pinatubo exploded in 1991, spewing millions of tons of sulfur into the atmosphere, global temperatures dropped 1°F in less than a year. (Another method would involve ships spraying seawater into the sky, brightening marine clouds and enhancing their solar reflexivity.) A report by David Keith—a geoengineering researcher at Harvard University—found that a small fleet of specially designed airplanes could inject 1 million tons of sulfur aerosols into the atmosphere at a cost of $1 billion to $2 billion a year, and that it would offset more than half the global warming that's already occurred. Blocking sunlight seems much less expensive—and perhaps politically realistic—than cutting carbon.

But if SRM geoengineering seems too good to be true, it may well be. Even if it's successful, geoengineering would have to be done continuously if carbon emissions kept rising—and if it stopped, because of a war or economic crisis, there could be a bounce-back effect that could cook the planet in a matter of years. Geoengineering wouldn't offset some of the other impacts of high carbon emissions, like ocean acidification. There would be side effects, some of which are known—like increased acid rain from all that sulfur—and scarier unknown unknowns. Climate, after all, is far more than just temperatures, and it's difficult to predict what might happen to rainfall patterns should we try to artificially cool the planet.

If the earth were a patient, geoengineers would be like 19th-century doctors—their ignorance means they could easily do more harm than good. And if you think trying to craft a global deal to curb carbon emissions has been tricky, trying to get 192 nations to agree on exactly how we should try to fine-tune the climate could be all but impossible. That also leaves the worrying possibility that one country—particularly afflicted by climate change—could decide to unilaterally experiment with geoengineering. "You are not going to get a reasonable global dialogue about a geoengineering research program," says Victor.

Still, the possible consequences of unchecked global warming are so scary that it would be smart to ex-

FROM LEFT: KAREN MOSKOWITZ—CORBIS; INTELLECTUAL VENTURES (3)

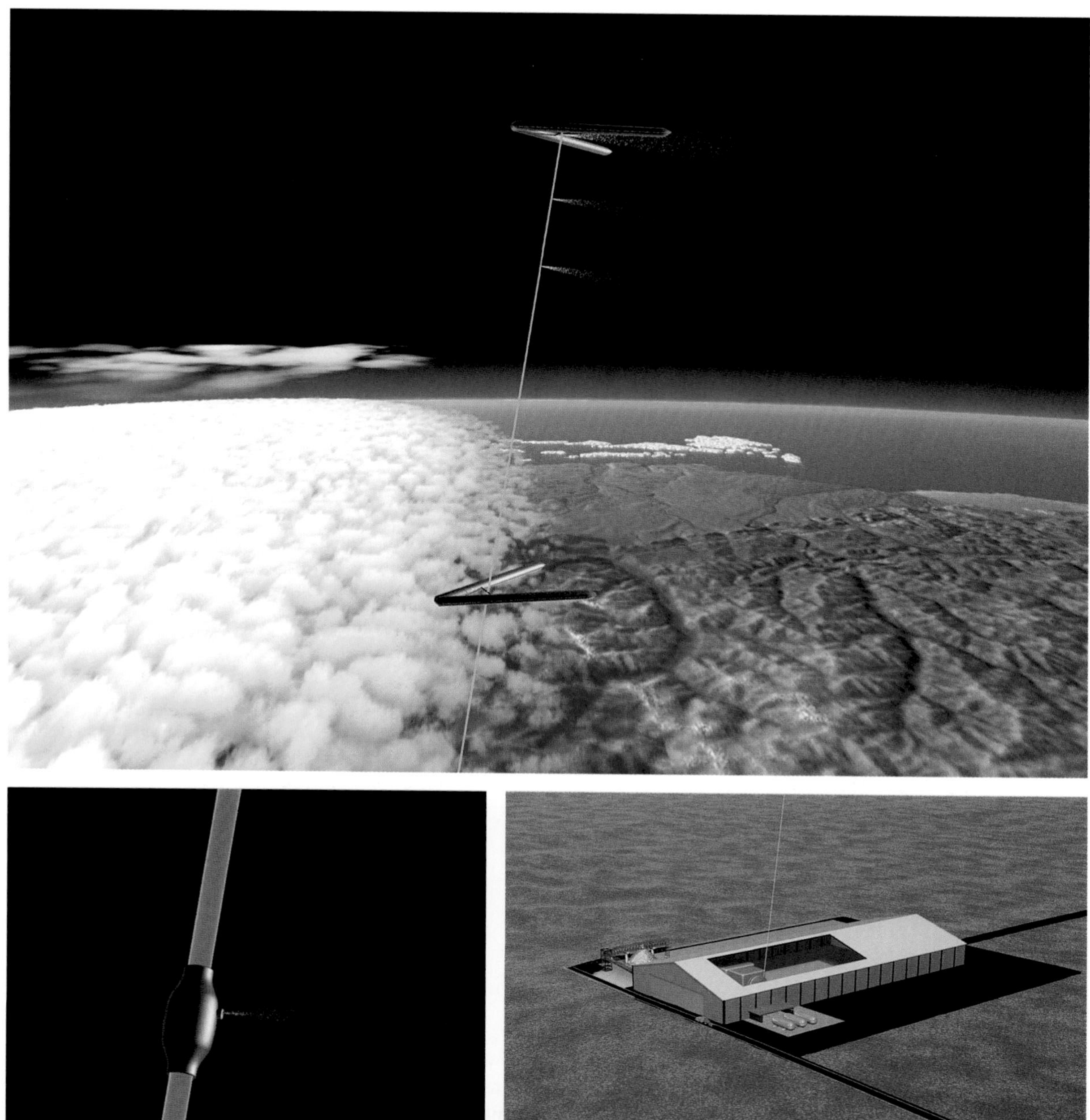

An illustration of the StratoShield, top, which would block sunlight from reaching the planet's surface, cooling the climate. Nozzles on the StatoShield, bottom left, would release sulfur particles, and base stations, bottom right, would control the shield.

plore geoengineering, so that at least we have an idea what will happen when we push the doomsday button. But little official work is being done outside computer models, and while a few tycoons like Bill Gates have funded small-scale goengineering research, no government has done so yet. The legal picture is murky—in 2011 the Convention on Biological Diversity (CBD) essentially banned geoengineering research except for small-scale trials, but no one knows exactly what qualifies and what doesn't. (And in any case, the U.S. isn't a party to the CBD.) Scientists worry that in the absence of clear rules and government funding, geoengineering might fall to rich individuals or rogue nations who aren't bound by ethics. "There is a race between responsible geoengineering research and those who will do it unilaterally," says Victor. "I'm worried countries could go off and just do things on their own." The right approach would be to research geoengineering in a fully open manner—and pray we never need to use it.

Making Carbon Pay Off

Can enough CO_2 be removed from the atmosphere to slow down or even reverse global warming? Four companies that have built machines to extract carbon from the air are betting big that the answer is yes.

By Marc Gunther

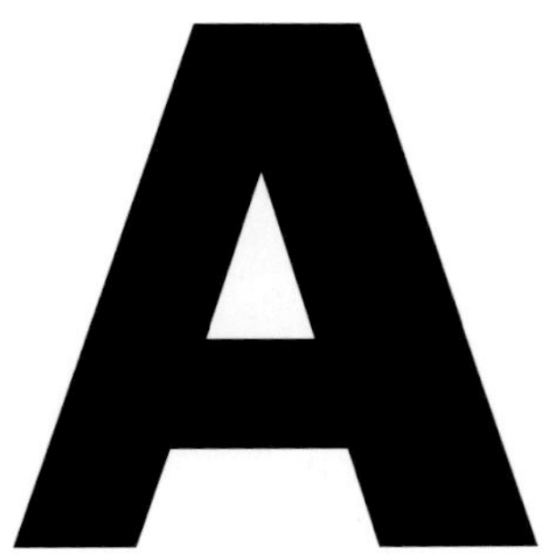

T HIS JOB AT THE LOS Alamos National Laboratory during the 1990s, German-born physicist Klaus Lackner grew worried about the threat of climate change. So when his daughter Claire asked for help coming up with an 8th grade science project, he asked her: "Why don't you pull CO_2 out of the air?"

Chemical engineers know that carbon dioxide, an acidic gas, will bind with sodium hydroxide, a base, to make carbonates. That's essentially how CO_2 is removed from the air inside of submarines or spaceships as the gas builds up from the breathing of the people inside. Claire accomplished the feat by filling a test tube with a solution of sodium hydroxide, buying a fish-tank pump from a pet store, and running air through the test tube all night.

"I was surprised that she pulled this off as well as she did," Lackner recalls, "which made me feel that it could be easier than I thought." Duly inspired, Lackner set out to design a machine to capture carbon dioxide from the air on a much larger scale, a quest that has obsessed him ever since.

But can enough carbon dioxide be removed from the

Marc Gunther *is a contributing editor at* Fortune *and the author of* Suck It Up: How Capturing Carbon From the Air Can Help Solve the Climate Crisis, *published in 2012 as an Amazon Kindle Single.*

CARBON ENGINEERING

Air contactors, like the one in this artist's rendering, would remove large quantities of carbon (CO_2) from the atmosphere.

atmosphere to slow down global warming—or even reverse it? And, if it can, who's going to pay for it? Most experts dismiss the idea as expensive and hopelessly impractical. But Lackner, who is now a professor at Columbia, is one of a number of iconoclastic scientists and entrepreneurs who believe that direct air capture of carbon dioxide will become a must-have technology if atmospheric concentrations of CO_2 continue to rise as expected. These scientists have launched startup companies and attracted well-to-do investors, including Microsoft co-founder Bill Gates and entertainment mogul Edgar Bronfman Jr. They have been persuaded that capturing CO_2 from the air is not merely worth investigating, but that it could become a real business—one that will be financed, improbably, by the oil industry that uses CO_2 to extract petroleum from deep under the earth.

Direct air capture of CO_2 is an appealing notion for at least three reasons. First, the technology capitalizes on the fact that greenhouse gases are global pollutants. Because the atmosphere functions as a conveyor belt, moving CO_2 to any place where carbon is drawn off, emissions spewing from a power plant in China or a tailpipe in Sao Paulo can be captured anywhere on earth. As a result, air capture may turn out to be the only practical way to capture emissions from cars, trucks, trains, ships, and planes, which will be with us so long as fossil fuels are used in transportation. Second, at least in theory, air capture can dial down CO_2 concentrations after they exceed safe thresholds, as some scientists say they already have. Build enough machines to capture CO_2, and we can have whatever levels of CO_2 in the atmosphere—and therefore whatever climate—we want. Third, and most

Global Thermostat co-founder Graciela Chichilnisky at the company's carbon-capturing pilot facility in Menlo Park, Calif. The company says it can capture CO2 for less than $50 a ton.

Proposed air-capture technology: The synthetic tree device, top, mimics the way trees suck in CO2; an air-capture module, bottom, sends the saleable CO2 it captures into a portable shipping container.

important, there's a market for CO_2, which could give the air-capture business an economic underpinning. A pipeline network already moves CO_2, most of it from natural sources, around the country, where it is sold to oil companies, industrial customers, firms that grow algae, and even the makers of fizzy drinks.

"The single largest waste product made by humanity is CO_2. Thirty gigatons a year," says Nathaniel "Ned" David, an entrepreneur and venture capitalist who is chief executive of Kilimanjaro Energy, a company created by Lackner back in 2004. "It's immensely valuable, and today we just blow it out the tailpipe. What if there were some way to actually capture it, use it, and make money?"

Although direct air capture is sometimes seen as a form of geoengineering—deliberate, planetary-scale actions to cool the Earth—it's actually quite different. Geoengineering is risky, imperfect, and controversial, and it raises thorny questions about who gets to decide if and when it is deployed. Its side effects are unknown. What's more, technologies that would reflect the sun's rays, which are the most promising approach to geoengineering, addresses one symptom of the climate problem (warmer temperatures) but not the cause (rising atmospheric concentrations of CO_2). Partly as a result, geoengineering as a climate strategy is stuck. Governments have provided only token funds for research, and there's no business model to support it.

Carbon dioxide removal, by contrast, targets the cause of global warming, namely, rising concentrations of CO_2. (Higher CO_2 levels also acidify the ocean, threatening marine life, a problem that blocking the sun's rays won't address.) Direct CO_2 capture doesn't create global risks or raise governance issues, at least not until it scales way up. And, unlike geoengineering, it is being financed by the private market. Machines are being built, albeit on a small scale.

Four companies are trying to turn direct air capture of CO_2 into a business. Lackner's Kilimanjaro Energy was initially financed with $8 million from Gary Comer, the founder of Lands' End. An avid sailor and philanthropist, Comer grew concerned about climate change after he sailed a yacht through the normally ice-bound Northwest Passage with relative ease back in 2001. (Comer died in 2006.) Last year, Kilimanjaro raised another $3.5 million in venture capital. Another venture, Carbon Engineering, is led by David Keith, a physicist and climate scientist who has a joint appointment at the University of Calgary and at Harvard's Kennedy School. Bill Gates is an investor, as is his friend Jabe Blumenthal, a former Microsoft executive who designed the first version of Excel. (Their investments haven't been disclosed.) Global Thermostat was formed by two Columbia University professors—Peter Eisenberger, a physicist and founder of Columbia's Earth Institute, who once led global research and development for Exxon, and Graciela Chichilnisky, an economist and mathematician. Its biggest investor is Edgar Bronfman Jr., the Warner Music CEO and heir to the Seagram's fortune, who has put in $15 million. Finally, there's Climeworks, a venture-backed company based in Zurich founded by a couple of graduate students at a science and technology university there.

The biggest problem with direct air capture—and it's a doozy—is cost. Last year, a committee of the American Physical Society produced a 100-page technology assessment, called "Direct Air Capture of CO_2 with Chemicals," which estimated that air capture will cost "$600 or more per metric ton of CO_2." The report concluded: "Direct air capture is not currently an economically viable approach to mitigating climate change." Other experts put the cost even higher, at $1,000 a ton. To put that in a climate-change context, the typical American is responsible for nearly 20 tons a year, about six tons of which comes from driving a midsize-gasoline–powered car. If the estimates are right, it would cost $12,000 to $20,000 per person for Americans to clean up after themselves.

Backers say the costs will be much less. Global Thermostat, which opened a demonstration plant in 2010 at SRI International, a Silicon Valley research firm, says it will be able to capture CO_2 for well under $50 a ton. Carbon Engineering, the company backed by Gates, will aim to situate its first plants in places where there is cheap natural gas, cheap labor, cheap land, cheap construction costs, and, ideally, strong demand for CO_2. "If we can find all those at once," says David Keith, its chief executive, "we're printing money." Meanwhile, Ned David of Kilimanjaro says CO_2 capture could do for the oil business what hydrofracking has done for natural gas, unleashing vast amounts of fossil fuels that might otherwise remain in the ground. "A money gusher," he calls it. Others, including Boeing, have looked into the possibility of using direct air capture to produce CO_2, combine it with hydrogen, and make fuels at the military's Forward Operating Bases.

The other big challenge is scale. Capturing enough CO_2 to alter the climate means thinking big. After all, building the coal and gas plants, factories, cars, trucks, planes, and ships that have spewed roughly 1.1 trillion tons of CO_2 into the atmosphere since the start of the industrial revolution has cost trillions of dollars and taken more than a century. Air capture, David Keith says, "won't succeed with the 20 people working in my company, Peter's company, and Klaus's company. This needs thousands of engineers. If air capture is going to succeed, it's going to take industrial might." It will also take time, he says: "There's no way you can do a useful amount of CDR (carbon dioxide removal) in less than a third of a century or maybe half a century."

For the moment, enhanced oil recovery presents the best business opportunity for the direct air-capture start-ups. Eventually, though, they would like to make oil. Peter Eisenberger of Global Thermostat says it's possible to capture CO_2 from the air, extract hydrogen from water, and combine them to make renewable, low-carbon transportation fuels in a process powered by solar energy. "This has always been for me the holy grail even back when I was at Exxon in the last energy crisis," Eisenberger says. "It solves the energy security issue since everyone has water and CO_2 from air." Every country in the world could produce its own oil and burn it. The CO_2 would be recycled, like newspapers or aluminum cans.

This may sound far-fetched but recent history suggests that it's equally far-fetched to believe that countries, rich or poor, will agree to stop burning coal, oil, and natural gas. Technology breakthroughs—air travel, nuclear energy, landing a man on the moon, building a communications network that can dispatch information or entertainment instantly to anyone in the world—always seem futuristic, until they don't.

CLOCKWISE FROM TOP RIGHT: (C) KLAUS LACKNER 2009; GRT 2009; CHRIS SCHMAUCH—GLOBAL THERMOSTAT

Solutions

Climate change is one of the most difficult challenges we will ever face. That's why we're lucky that the smartest minds on the planet are hard at work on the problem. From new forms of energy to better efficiency, here are ideas that could light up the globe.

HENRIK SORENSEN—GETTY IMAGES

Top Green Tech Ideas

You may have to look hard, but some very smart companies are doing some very creative things when it comes to the environment.

Bloom Energy

FUEL CELLS

Sometimes high tech can start out low tech. Fuel cells are an old and basic technology; they generate electricity within a cell through the reaction of a fuel and an oxidant. Essentially they're a kind of chemical battery, but unlike batteries, they can't store electricity; fuel cells require an outside fuel source that has to be replenished over time. But their simplicity has also made them useful for certain purposes; NASA has long used hydrogen fuel cells to power its spacecraft.

Inventors have tried to use hydrogen fuel cells as a cleaner way to create electricity commercially. Honda and other car companies have made hydrogen fuel-cell–powered cars, for example, but they've always been limited by the cost. That's beginning to change, however, thanks to a California start-up called Bloom Energy. The company exploded onto the public scene in 2010 with the release of its Bloom Box, a system that uses fuel-cell technology to provide off-the-grid power. The Bloom Boxes—about half the size of a shipping container—use solid oxide fuel cells, which generate electricity by oxidizing natural gas. The technology has existed for awhile, but Bloom figured out how to carry out the reaction at a relatively low temperature, making the Bloom Boxes safe to use in corporate offices. This is exactly where they're being put to work now, by companies like Google and eBay that can use the lower carbon power as an off-the-grid backup to conventional grid electricity and as a way to reduce their own carbon footprint.

Bloom Energy's fuel cells could provide an inexpensive and clean backup power source for high-tech companies.

CloudBlue

RECYCLING E-WASTE

High-tech may have a clean image—all smooth-edged iPhones and liquid crystal displays—but the elements that go into phones and PCs can be polluting to the environment and dangerous to human health if incorrectly disposed of. So-called e-waste is the fastest-growing part of the solid waste stream, and some 20 to 50 million metric tons of it are thrown out every year.

CloudBlue, based in New Jersey, helps tech companies take care of their e-waste, arranging for direct pickup and processing, ensuring that valuable metals can be reused and recycled for future electronics. For customers like banks that have to worry about sensitive data that might be encoded on old computers, CloudBlue can also process the waste on-site. With all this, the company can ensure that no e-waste will ever end up in a landfill—or worse, poison a child in Africa or Asia.

CLOCKWISE FROM TOP RIGHT: RICHARD BERENHOLTZ—THE DURST ORGANIZATION; RENAULT; DON MASON—CORBIS

Better Place

ELECTRIC CARS

It's an article of faith among many environmentalists: The future will be electric. But how long is it going to take? Electric cars have been around since the dawn of the automobile—in fact, the technology hasn't changed all that much since Henry Ford's own electric Model-Ts. But the electric car lost out to gasoline-powered ones for good reasons: Gasoline carries a lot of power per gallon, while batteries never had the capacity to move cars very far. Even in the 1990s, with the introduction of improved electrics like GM's lamentedly discontinued EV1, battery-powered cars remained a fetish for those who value their carbon footprint over convenience. That's a nice shift in public attitude, but it's not a route to changing transportation as we know it

Times really have changed, though—and it's the tipping point for electric cars. Toyota has sold over a million Prius hybrids in the U.S. and GM's long-awaited Volt hybrid has finally arrived. The Japanese car company Nissan has gone one better with its all-electric Leaf—the one with the polar bear ads—as has Ford with its all-electric Focus. Smaller startups are experimenting with ultra-efficient electric cars, while the innovative company Better Place is installing networks of battery-charging stations in Israel for its own electric transportation system, with a subscription payment system modeled on the wireless industry. Electric cars still have a number of obstacles to overcome, and they won't make a huge dent in carbon emissions unless the grid itself is steadily cleaned up, but they are closing in on the competition.

Electric cars are just beginning to go mainstream.

Bank of America's new tower is state-of-the-art.

Serious Materials

GREEN BUILDING

Want your new building to stand out? Make it green. Green architecture has gone from a niche interest to an industry. Skyscrapers like the Bank of America headquarters in Manhattan are known for their energy efficiency, particularly their score on the Leadership in Energy and Environmental Design (LEED) scale. What MPG is for cars, LEED is for buildings. Sustainability has even become a point of competition among mega-mansion owners, with multimillionaires in California jostling to build the greenest house in America.

Much of green architecture comes from design, but smarter building materials can make a difference as well. Companies like Serious Energy produce highly efficient windows, insulation and other building features that reduce the amount of heat lost to the outside. Built right, "passive houses" can even be so energy efficient that they require no outside heat at all, bringing energy bills close to zero.

Most wind turbines are huge, but WindTronics has developed rooftop devices for harnessing wind power.

WindTronics

ROOFTOP WIND POWER

If you want to provide off-the-grid power for your own home, there's only been one solution: solar panels. Wind power is usually deployed on a utility-scale, in vast farms of mighty turbines that feed directly into the grid. Solar has always been the choice for homeowners who want to stop paying electricity bills and start generating their own juice.

But if wind can do big, it can also do small—and it does rooftops as well. The startup WindTronics is developing mini-wind turbines that can be installed on any flat roof, either alone or in larger arrays. Each turbine measures about 6 feet in diameter and looks like a large, circular window fan, but it can generate an average of 1,500 kilowatt hours a year, with more or less depending on wind strength. And unlike utility-scale turbines, the WindTronic turbine contains no rotating gearbox to generate electricity, and is thus much quieter. In an ordinary wind turbine, the blades move the gears, the gears turn a generator, and the generator creates electricity. With a WindTronics model, the blades are equipped with magnets at the tips and are enclosed in a wheel that contains coiled copper, so the entire turbine is an electric generator.

eSolar has an innovative way to capture the energy of the sun.

eSolar

SOLAR TOWER

Bill Gross at eSolar thinks that he can improve on solar power. Instead of PV panels, eSolar uses vertical mirrored towers that concentrate sunlight on a ground target. Using software that Gross helped write himself—he was an Internet entrepreneur before breaking into alternative power—the mirrors perfectly track the sun as it crosses the sky, maximizing the amount of electricity that can be produced. The result is a relatively compact but power utility-scale plant that gets the most out of that free source of energy called the sun.

Itron

SMART METERS

Our electrical devices may be 21st century, but the electrical grid we plug them into is strictly 20th. Improving the grid is going to be a vital part of helping clean energy scale up and the first installment on a smarter grid will be smarter meters. Right now the electric meter in your home tells you—and the electric company—only the most basic information. The majority of utilities won't even know that homes have lost power in a blackout until enough annoyed customers call them. But smart meters connected to a network can relay that sort of information instantly, giving utilities and customers alike a real-time picture of how much power is being used at any given moment. And as new appliances are networked into smart meters—like the kind offered by Itron—we'll be able to use them much more efficiently. By smoothing out the electricity demand curves, smart meters can help utilities get more out of the power plants they already have.

CLOCKWISE FROM TOP LEFT: FREDERIC A REINECKE—WINDTRONICS; NOVACEM; STEVE KING—ESOLAR

Novacem

GREEN CONCRETE

Cement making accounts for around 5% of global carbon emissions. The good news is that there are enormous carbon savings that could be realized by making cement production more energy efficient. Some firms are reformulating the products used to waterproof concrete in a way that saves energy. The startup Novacem is going further, working on a new cement production method that would actually absorb more CO_2 than it releases, by substituting carbon-rich limestone with magnesium silicates that contain no stored carbon. As the cement hardens, CO_2 in the air actually reacts to make solid carbonates that strengthen the cement while holding onto the gas. If Novacem's process can be applied on a commercial scale, concrete could become carbon negative.

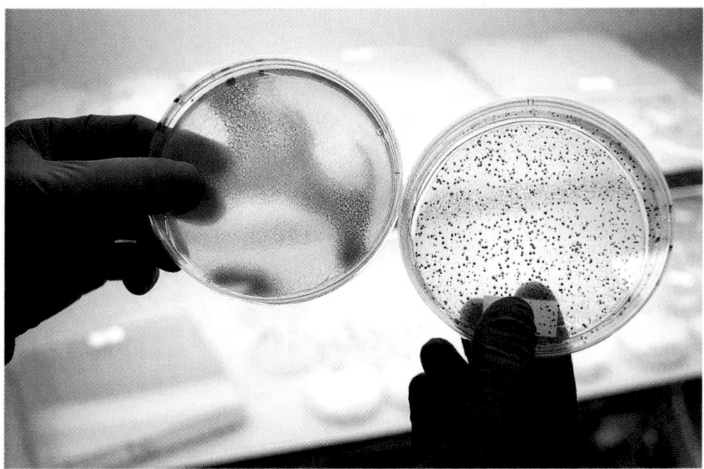

Biofuel from algae is more efficient and greener than corn ethanol.

Sapphire Energy

ALGAE BIOFUEL

Thanks largely to generous government subsidies, the U.S. produced over 11 billion gallons of corn ethanol in 2011. That was enough to displace the need for 364 million barrels of oil, but study after study has shown that high levels of corn ethanol production simply aren't sustainable. Corn that could go to feed the world instead feeds our cars—and not very efficiently.

But that doesn't mean biofuels can't play a major role in a greener U.S. energy policy—they just have to be the right kind. One of the best options on the horizon is biofuel made from algae, which counters a lot of the problems with corn ethanol. Algae don't need farmland to grow: Tanks will do the job anywhere there is spare land and sunshine. Algae also grow much faster than traditional crops, and the micro-organisms may be able to use wastewater, rather than fresh supplies. Startups like Sapphire Energy are passing the pilot phase and nearing development.

CarbonZero Project

BIOCHAR

Given the scale of the climate challenge, everyone wants to find a silver bullet, a way to cut carbon emissions quickly and cheaply. Until someone perfects cold fusion, a cleaner economy will require a portfolio of new and innovative technologies, each playing its part. But that doesn't mean there aren't shortcuts on the road to zero carbon. Here's a simple one: biochar.

Plants absorb carbon dioxide as long as they're alive, but once they're cut down or burned, that carbon is released back into the atmosphere. Keeping trees standing—especially in tropical areas—is one way to save that carbon. But if plants are cut down, perhaps for agriculture, and you burn the residue in a controlled, low-oxygen atmosphere—a simple process called pyrolysis—you can create charcoal, a stable and solid form of carbon. If you then mix the biochar with certain soils, you can also reduce the amount of methane and nitrous oxide, both of them greenhouse gases, that the soil would naturally release. The result is a two-for-one carbon-cutting special, and the potential is tremendous. A study in *Nature Geoscience* found that biochar could offset 12% of global carbon emissions. The challenge is that biochar has little value on its own, so there's not much business case for making the product right now. That's one more reason a carbon price would be so useful.

Verdant

TIDAL POWER

Tides are the winds of the oceans, generating a tremendous amount of kinetic energy that can be tapped with the right kind of technology. In fact, tides might be better than wind, since they're much more predictable. And while the best wind resources tend to be located far from major population centers, most of the big cities around the world are located right next to the water. The problem has always been that building turbines is significantly more expensive underwater than on land.

That's still the case, but tidal power is slowly beginning to gain acceptance. The technology works the same way a wind turbine does: The steady movement in and out of the tides turns an underwater turbine, which generates electricity. And as with wind, there are some parts of the world that are particularly rich in tidal potential, like the Bay of Fundy in Canada, home to some of the most intense tides on the planet. New York City's tides are a lot calmer, but the city does have potential for tidal power.

Tidal power is predictable, but costly.

CLOCKWISE FROM TOP LEFT: DAVID MAUNG—BLOOMBERG/GETTY IMAGES; LEONARD GERTZ—CORBIS; VERDANT POWER

Joule Biotechnologies

ARTIFICIAL PHOTOSYNTHESIS

As smart as human beings can be, nature almost always does it better—possibly because nature has had hundreds of millions of years to get it right. Take photosynthesis for example. Plants with green leaves are able to capture the sun's energy and turn it into useful chemical fuel in a process that is much, much more efficient than our best photovoltaic solar panels. We can save nature by copying nature.

That's why there are a number of scientists working on creating artificial photosynthesis; it was even a major plot point in *Solar,* the English writer Ian McEwan's global-warming–themed novel. Daniel Nocera, an energy expert at the Massachusetts Institute of Technology and a Time 100 awardee, is pushing a form of artificial photosynthesis that would create electricity that would then be harnessed to produce hydrogen for use in fuel cells. That's only one way to control photosynthesis, but already startups like Joule Biotechnologies are looking for ways to take it commercial. The future has to be solar-powered; the question will be how best to tap into that free source of energy. The trees might have the best idea.

ESSAY

Children of a Hot Planet

By Bill McKibben

Each spring I seem to end up giving two or three commencement addresses, so I get to watch as a lot of well-scrubbed and well-educated young people head out into what we call the real world. I've got a daughter of my own, who with any luck should be graduating college herself in the next few years. And when I look at them, or when I think of her, I confess that I shudder a little. For most of the history of the Republic, parents have been able to tell themselves that they're leaving their children a better world. Atmospheric science would indicate that we're leaving ours an unprecedented mess.

I wrote one of the first books for a general audience about climate change, back in the 1980s when I was in my twenties. In the subsequent quarter-century, two things have changed. One, the science has grown darker: We've come to understand that even small changes in temperature (barely a degree and a half F., averaged globally) can melt vast areas of ice, and trigger Biblical-scale flood and drought. That's bad.

The second thing's worse, however: Our political system has so far failed this greatest of tests. In Washington we've had a 20-year bipartisan effort to accomplish essentially nothing; globally the picture is at least as bleak, in part because many developing nations, led by China, have turned into carbon-spewing regimes based on our model.

If we continue down this path, the scientists offer little solace. The head of the International Energy Agency said in the spring of 2012 that we're on pace to increase the planet's temperature by nearly 11 degrees, much of it in the lifetime of people now being born. If anything close to that happens, the result will be a science-fiction kind of world, one in which the choices of a few decades will echo for geological time.

We needn't go down that road, however. The same quarter-century has seen steady progress in the development of alternative energy. I'm typing these words on a laptop powered by the sunlight falling on my roof; there are prosperous European nations already drawing a quarter of their power from the wind. New "net zero" houses return energy to the grid; some of the most fashionable cities on the planet have retooled for pedal power. We know how to do most of what we need to do—the question is whether we can make the large-scale shift quickly enough to matter.

My guess is that the key will lie less in building new technologies than in building movements. At the moment, the political power of the fossil-fuel industry thwarts real change—they're the richest enterprise in human history. But this is not a typical political fight, with opponents divided by ideology. Instead of Republicans vs. Democrats or industry vs. environmentalists, the real fight boils down to human beings against physics and chemistry.

Bill McKibben *is the Schumann Distinguished Scholar at Middlebury College and a co-founder of the environmental group 350.org. He is the author of more than a dozen books about nature and the environment.*

TAKUYA UROKU—ANYONE/AMANAIMAGES/CORBIS

Physics and chemistry will win that fight. Indeed, the most recent polling data shows more and more Americans are concerned about climate change, precisely because they can see the freak weather it has begun to trigger. The year 2012 witnessed heat waves so bizarre that the mercury touched 94°F in the Dakotas in the winter. Eventually we'll wake up.

When we do, our work will be cut out for us. We'll need to spur the pace of energy innovation, just as we needed to learn to build tanks and planes at an unprecedented pace during World War II. And we'll need to make some short-term economic sacrifices—in particular, we'll need to put a price on carbon to reflect the damage it causes. That will sting: there's no painless cure. But it will offer us a future, and in some ways a sweet one. Imagine a world where your energy doesn't come from a few regional power stations, but from a million interconnected solar panels on a million neighboring roofs. Imagine, that is, a grid that looks less like old broadcast television and more like the Internet.

We can try to hang on to the present for a few more decades, pretending that we can avoid the future scientists assure us is coming. Or we can rise to the occasion. Imagine this moment as a test of whether the big brain was a good adaptation or not: Can it foresee the trouble we're in? Is it connected to a heart big enough to propel the shift we need? The generation now coming of age will bear most of the cost either way, and the exam they need to pass is the one we've largely failed. The one about leaving behind a better planet.

20 Things You Can Do

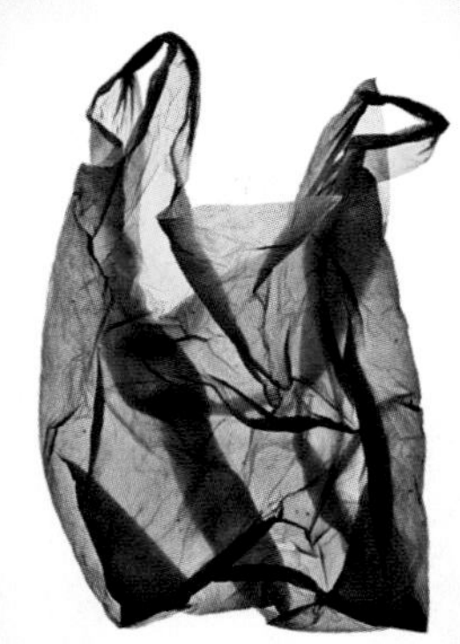

1. Check the Tires—and the Car.
Tuning up your engine can improve gas mileage by at least 4%, if not more. Keeping your tires properly inflated can increase mileage by more than 3%.

2. Carpool!
Nearly 80% of people drive to work alone, but it doesn't have to be that way. Carpooling can help you reduce carbon emissions and save on gas.

3. Turn Off Your Office Lights.
Far too many of us simply leave our lights on when we leave work for the night. That's wasteful—we should aim to ensure that off-peak energy is one-fifth of peak use.

4. Shut Off Your Computer.
Seventy-five percent of the electricity used by home electronics is consumed in standby mode. Shutting off your PC can reduce carbon emissions significantly.

5. Forgo the Plastic Bag.
Every year more than 500 billion plastic bags are distributed, and each needs a little bit of oil. Switch to reusable bags to save emissions.

6. Buy Energy Star.
The rating system Energy Star can help you find more efficient, cheaper appliances. It can help you reduce your power bill by as much as 30%.

7. Go Vegetarian—Sometimes.
The meat industry—and especially beef—is responsible for a huge chunk of global carbon emissions, not to mention its massive use of water and grain. Go vegetarian at least once a week and it will make a difference.

8. Get an Energy Audit.
A home energy audit, which you can usually get for free, can tell you how much power your household uses and how to reduce it. Take advantage of it—you can save carbon and energy.

9. Open a Window.
Most of the carbon emissions produced in a home come from heating and cooling it. For natural cooling, open windows at night and close them in the day.

10. Pay Your Bills Online.
If every U.S. home owner viewed and paid bills online, rather than waiting for the snail mail, the carbon savings could exceed 2 million tons a year.

11. Move to the City.
Environmentalists may embrace the countryside, but the greenest place in the U.S. is Manhattan. Dense urban living reduces energy wasted on transportation and building.

12. Ride the Bus.
Transport accounts for more than 30% of U.S. carbon emissions, and one of the best ways to reduce that footprint is through public transport. There's a lot of room for improvement—over 80% of U.S. trips are by car.

13. Telecommute.
A study of Boeing found its 80,000 workers in Puget Sound traveled the equivalent of 85 times around the Earth on their daily commutes. Working from home would help.

14. Hang Up a Clothesline.
Sixty percent of the energy associated with clothing is spent on washing and drying. Air dry to save carbon.

15. Downsize Your Home.
Houses in the U.S. have bulked up in recent decades, using more and more energy, even as family sizes have shrunk. A smaller home will cost less to heat and cool—and clean.

16. Change Your Lightbulbs.
It's a cliché, but true—the quickest way to cut energy costs is to trade your incandescent bulbs for compact fluorescent. You'll save emissions—and won't need to replace your bulbs so often.

17. Go Hybrid—or Electric.
Hybrid cars are no longer unusual—the Toyota Prius is one of the best-selling models in the U.S. If you're really ambitious, try an electric like the Chevy Volt.

18. Take Showers, Not Baths.
The energy required to heat water is a major part of your home carbon footprint—and surprisingly, baths require a lot more water than showers. Switch up to save water and energy.

19. Go Solar.
Installing solar panels used to be a pain, but today companies like Solarcity of California offer one-stop shops for solar installation. Take advantage of it—solar now can help you cut your utility bills for years, and government subsidies can reduce your upfront costs.

20. Vote!
Personal actions are important, but only legislation and other political activity will really help save the planet. Support politicians who are willing to make the tough choices to deal with climate change.

FROM LEFT: ANGUS FERGUSSON—RADIUS IMAGES/CORBIS; STEVE GALLAGHER—CULTURA/CORBIS